Ending F-22A Production

Costs and Industrial Base Implications of Alternative Options

Obaid Younossi, Kevin Brancato, John C. Graser, Thomas Light, Rena Rudavsky, Jerry M. Sollinger

Prepared for the United States Air Force

Approved for public release; distribution unlimited

The research described in this report was sponsored by the United States Air Force under Contract FA7014-06-C-0001. Further information may be obtained from the Strategic Planning Division, Directorate of Plans, Hq USAF.

Library of Congress Cataloging-in-Publication Data

Ending F-22A production : costs and industrial base implications of alternative options / Obaid Younossi ... [et al.].
p. cm.
Includes bibliographical references.
ISBN 978-0-8330-4649-9 (pbk. : alk. paper)
1. F/A-22 (Jet fighter plane)—Costs 2. F/A-22 (Jet fighter plane)—Design and construction—Evaluation. 3. United States. Air Force—Procurement—Evaluation. I. Younossi, Obaid.

UG1242.F5E53 2010
338.4'76237464—dc22

2009027954

Published 2010 by the RAND Corporation
1776 Main Street, P.O. Box 2138, Santa Monica, CA 90407-2138
1200 South Hayes Street, Arlington, VA 22202-5050
4570 Fifth Avenue, Suite 600, Pittsburgh, PA 15213-2665
RAND URL: http://www.rand.org/
To order RAND documents or to obtain additional information, contact
Distribution Services: Telephone: (310) 451-7002;
Fax: (310) 451-6915; Email: order@rand.org

Preface

The F-22A Raptor is the most advanced fighter aircraft in the world. It is currently in production, with the last of 187 authorized aircraft to be procured by May 2010. In April 2009, the Department of Defense decided to terminate production of the aircraft, with the production line closed down after the last aircraft delivery. In advance of this decision, the Air Force asked RAND Project AIR FORCE to identify the associated costs and implications of various shutdown options on the industrial base. This monograph describes the findings of this research.

Related RAND Corporation documents include the following:

- *Reconstituting a Production Capability: Past Experience, Restart Criteria, and Suggested Policies,* by John Birkler, Joseph P. Large, Giles K. Smith, and Fred Timson (MR-273-ACQ), 1993.
- *F-22A Multiyear Procurement Program: An Assessment of Cost Savings,* by Obaid Younossi, Mark V. Arena, Kevin Brancato, John C. Graser, Benjamin W. Goldsmith, Mark A. Lorell, Fred Timson, and Jerry M. Sollinger (MG-664-OSD), 2007.

This monograph was prepared as part of a study entitled "The Future of F-22A Production." It was sponsored by Maj Gen Jeffrey Riemer, former Air Force Program Executive Officer (PEO) for F-22A, and conducted within the Resource Management Program of RAND Project AIR FORCE. The project monitors were Lt Col Elisa Coyne, F-22A PEO, and Doug Mangen, F-22A program office. The monograph should interest those involved in the acquisition of defense systems.

The data collection and analysis for this study were completed in March 2008, and the final briefing was presented to the Air Force in April 2008. Subsequent review by the Air Force in April 2009 suggests a substantial increase in the number of tools required for restarting production at a later date. Other data inputs may have changed as well. We opted not to use the new estimates since we are unable to validate them. In addition, in this analysis we utilize the program of record at the time of this research (183 aircraft), although the program of record had grown to 187 by the time of printing.

RAND Project AIR FORCE

RAND Project AIR FORCE (PAF), a division of the RAND Corporation, is the U.S. Air Force's federally funded research and development center for studies and analyses. PAF provides the Air Force with independent analyses of policy alternatives affecting the development, employment, combat readiness, and support of current and future aerospace forces. Research is conducted in four programs: Force Modernization and Employment; Manpower, Personnel, and Training; Resource Management; and Strategy and Doctrine.

Additional information about PAF is available on our Web site: http://www.rand.org/paf/

Contents

Figures

Tables

Summary

Background

The U.S. Air Force's F-22A Raptor is the world's most advanced fighter aircraft. Currently, Congress has authorized the procurement of 187 F-22As. The final production funding for the program of record is in fiscal year (FY) 2009, with the last delivery about two years later. After this, the Air Force will continue with contracts that provide for some modernization and sustainment work but no new production. Since Congress has prohibited the sale of the F-22A to other countries, there are no options to keep the production line active. RAND Project AIR FORCE was asked to evaluate various scenarios for the time that full-rate production is no longer an option.

Purpose

This monograph explores four options for maintaining a future F-22A industrial capability after the last aircraft is delivered. Descriptions of these options follow.[1]

Shutdown

In this option, the production line would close permanently once the last aircraft is delivered. Tools, special test equipment, and assembly fixtures would be either disposed of or reallocated to other activities, such as F-22A sustainment and modernization or other programs. Professional employees and production workers would be reassigned or let

[1] These options were selected prior to the decision to terminate F-22A production.

go. Production facilities would be reconfigured and reassigned to other work or vacated. This alternative does not include the cost of retaining tools that may be required for future performance upgrades or a service life extension program (see pp. 3, 14–20).

Shutdown and Restart

In this option, the production line would be shut down in a way that would facilitate restart. We assumed a production gap of two years.[2] This option requires significant planning for storage and maintenance of most tools, test equipment, technical information, and production facilities. Further, due to the complex nature of the F-22A design and manufacture, a core of highly trained engineers and production workers, as well as the industrial base capability, must be retained to make a smooth production restart possible (see pp. 4, 9, 21–32).

Warm Production

Under this option, a small number of aircraft would be produced until enough funds were available to return to full-rate production. All the means of production would remain in place, albeit used inefficiently. The low level of production would require a reduction of force, mothballing of some equipment and tooling, and subsequent rehiring and training of workers (see p. 4).

Continued Production

Under this option, production would continue at the current rate (see p. 5).

---

[2] Birkler et al., 1993, examined the effect of 11 aircraft, helicopter, and missile programs and concluded that the length of the gap did not correlate with recurring production costs at restart.

Results

Costs

Table S.1 shows the results of our analysis. It presents the aggregate costs of termination, hiatus, restart, and production, across the four options, utilizing common cost comparisons. Costs are in constant FY 2008 dollars, except for the final column, which lists total costs in then-year (TY) dollars, reflecting the Air Force's interest in the effect on future budget requirements.

Each production option compares the costs of an additional 75 aircraft.[3] The second column contains the sums of all hiatus, restart, and termination costs for each scenario. The third column contains flyaway unit costs, which are the sum of the target price curve (TPC),[4] propulsion, and other elements of flyaway cost, divided by 75; these range from a low of $139 million for the Continued option, to $154 million for Warm Production, to $179 million for Shutdown and Restart. The fourth column is the average unit cost (AUC), which includes all costs—hiatus, restart, termination, procurement, fly-

Table S.1
Total Cost, by Scenario (FY08$)

Options (2010–2016)	Hiatus, Restart, and Termination Costs ($million)	Flyaway Unit Costs ($million)	Average Unit Cost ($million)	Total Cost ($billion)	Total Cost (TY $billion)
Shutdown	79	—	—	0.4	0.5
Shutdown and Restart	513	179	227	17.0	19.2
Warm Production	111	154	213	16.0	17.8
Continued Production	79	139	173	13.0	13.8

[3] Each production option assumes a total of 75 production aircraft after Lot 9. This assumption was made to keep the Continued Production lot sizes comparable to previous F-22 production lot sizes and the total quantity comparable among all production options. Also, contractor-developed rates and factors did not allow for analysis beyond FY 2016.

[4] Details of what the TPC includes are discussed in Chapter Two.

away, and below-the-line[5]—divided by 75.[6] The fifth column contains total cost, which is AUC multiplied by 75, except for the Shutdown scenario, in which the $0.4 billion cannot be spread over any aircraft.[7] The final column lists total cost in TY dollars. Note that the Shutdown and Restart and the Warm Production options both impose a per-aircraft cost penalty (see pp. 34–52).[8]

From a total cost perspective, Shutdown has the lowest cost; Shutdown and Restart, the highest. Continued Production or keeping the production base warm with low-rate production falls in between.

Sustainment, Modernization, Technical Data Package, Contract Closeout

In terms of costs to the F-22A program, Shutdown would produce no additional costs above the baseline program of record in program sustainment, modernization, technical data package, and contract closeout. This does not imply that all these efforts have been fully funded in the Air Force Program Objective Memorandum or budget; rather, it indicates that requirements should not change due to a decision not to continue production after Lot 9, the last lot of the multiyear buy (see pp. 55–66).

Effect on the Industrial Base

One concern over shutting down the production line either permanently or for some interval has to do with the effect it would have

---

[5] Below-the-line costs are discussed in detail in Chapter Two.

[6] Note that AUC is for the next 75 units only. This should be clearly distinguished from average procurement unit cost and program acquisition unit cost, both of which are calculations of cost of all units procured since the start of a production run, which would in this case include the 183 F-22As already procured (the program of record at the time of this analysis) plus the next 75.

[7] Roughly $330 million (FY08$) of the total cost of each scenario represents program support sustainment activities that are currently funded in the production budget. These costs will transfer to the sustainment budget during a production gap and after termination. Although these sustainment costs are included in all scenarios for comparability, they do not represent additional or unforeseen costs of terminating production.

[8] No matter what option is chosen, at some point the production will end, so there will be shutdown costs for every option.

on the aircraft industrial base. The F-22A is a complex aircraft that requires a range of highly technical skills. These skills reside in many firms that contract with the government or with a prime contractor to build or assemble portions of the aircraft. If the Air Force stops building F-22As, some of these contractors must work on other aircraft or turn their attention to other areas, perhaps even leaving the field of military aircraft production. Once a particular contractor enters a different field, it may be difficult to attract it back to aircraft production if a future need arises for its services. To gauge the effect of a shutdown on the contractor workforce, RAND surveyed the prime contractors to understand their perception of how a production gap may affect their vendors. The airframe prime contractors were reluctant to allow RAND to directly survey their vendors because they ascertained that the survey would make future negotiations between the prime contractor and the vendors difficult. Note that we examined only the effect of Shutdown and Restart because the other options either terminated production or continued it at some level.

Our analysis of the responses suggests that the prime contractors expect that only a few of their vendors would go out of business as a result of a two-year production gap. The prime contractor surveys indicated that about 20 percent of their subcontractors were at high risk (50 percent chance) of having issues that might compromise their availability. The biggest concerns in other areas were the unavailability of workers with security clearances for specialized skills or processes, general skilled labor, or facilities.

However, other issues are likely to hinder the program's restart capabilities. These issues include requalifying the vendor base, as well as concerns over the availability of skilled labor, processes, facilities, and tooling used by firms supporting F-22A production. All these issues are likely to affect suppliers' ability to provide the same parts when production starts again as they provided when it closed down. Some parts will be technologically obsolete; in other cases, the facilities dedicated to making those parts will have closed down or will have been diverted to manufacturing other parts (see pp. 67–76).

Acknowledgments

The authors are grateful to Maj Gen Jeffrey Riemer (retired), former F-22 PEO, for sponsoring this project and opening doors at the F-22 Systems Program Office (SPO) and contractors. We thank Lt Col Elisa Coyne, from the F-22A PEO office for coordinating contractor visits and briefings to the Air Force leadership, Office of Secretary of Defense, and congressional staff. We also thank Doug Mangen and Ken Birkofer from the F-22 SPO for providing F-22 historical cost information and much more.

Additionally, we would like to thank Jack Twedell and Mark Byars from Lockheed Martin; Bob Jenkins and Bill Cribb from Boeing; and Jeff Zotti and Mike Garcia from Pratt & Whitney for hosting meetings at their facilities, providing data, and visiting RAND on a number of occasions. Without their and their team's data and insights, this study would not have been possible.

We are also grateful to our RAND colleagues Edward Keating and Fred Timson for carefully reviewing the draft manuscript and suggesting many substantive changes that enormously helped the quality of this report.

Other RAND colleagues who provided helpful insights and encouragement during the study are Andrew Hoehn, Laura Baldwin, Cynthia Cook, Natalie Crawford, and Mark Arena.

Finally, we thank Chris Sharon for his research assistance, Brian Grady for his research and administrative support, and Miriam Polon for editing this monograph.

Abbreviations

ALC	air logistics center
APUC	average procurement unit cost
AUC	average unit cost
BAE	British Aerospace
COA	course of action
CRI	cost reduction initiative
DMS	diminishing manufacturing sources
EMD	engineering and manufacturing development
FAR	Federal Acquisition Regulation
FASTeR	Follow-on Agile Support To the Raptor
FOC	full operational capability
FS&T	field support and training
FY	fiscal year
GFE	government furnished equipment
JSF	Joint Strike Fighter
MUN	minimum unit number
MYP	multiyear procurement

OEM	original equipment manufacturer
OGC	other government costs
PAF	RAND Project AIR FORCE
PALS	Performance-based Agile Logistics Support
PAUC	program acquisition unit cost
PEO	program executive officer
PIP	Producibility Improvement Program
POM	Program Objective Memorandum
PRTV	production representative test vehicle
PSAS	Program Support Annual Sustaining
PSO	Program Support Other
PSP	Program Support Producibility
PSS	Program Support Sustaining
RUP	retained unit percentage
SPaRE	Sustainment Program for the Raptor Engine
TPC	target price curve
TY	then-year
USAF	United States Air Force

CHAPTER ONE

Introduction

Background

Characteristics of the F-22A

The F-22A Raptor was designed as a two-engine air-superiority fighter with both air-to-air and air-to-ground capabilities. Unlike other fighter aircraft in service, its advanced avionics and stealth capability reduce enemy opportunities to track and engage the aircraft with missiles (U.S. Air Force, 2008). Its capacity to enter supercruise flight allows the aircraft maximum speed and range while avoiding the use of fuel-consuming afterburners. Two advanced turbofan F119 engines with thrust vectoring capability enhance the aircraft's maneuverability. These design measures make the F-22A unique in its capacity to track, identify, shoot, and kill advanced air-to-air threats before being detected (U.S. Air Force, 2008). So far, the U.S. Congress has prohibited international sale or manufacturing collaboration of the F-22A, keeping the United States Air Force (USAF) as the sole operator of this aircraft. Some allied countries have expressed interest in this aircraft's capabilities. For example, Japan assessed the F-22A as the most logical aircraft to counteract future threats due to its multi-engine and stealth capabilities (Bennett, 2007).

History of the F-22A Program

In the 1980s, USAF officials identified the need for an Advanced Tactical Fighter program to replace the fourth-generation F-15, which had been designed in the early 1970s. On April 23, 1991, a joint Lockheed Martin/Boeing team won the air vehicle development contract. Pratt

& Whitney won the engine development contract with its innovative F119 design. The engines are provided to Lockheed Martin as government furnished equipment (GFE). The F-22A Raptor reached the first flight milestone on September 7, 1997. On December 15, 2005, the F-22A achieved initial operational capability. Two years later, on December 12, 2007, the F-22A achieved full operational capability (FOC), making the aircraft ready for global engagement for the first time.

Aircraft Production

The aircraft's subassemblies and parts are manufactured in several places across the country. The prime contractors include Lockheed Martin, Boeing, and Pratt & Whitney. Lockheed Martin produces the mid-fuselage in Fort Worth, Texas, and the forward fuselage and final assembly in Marietta, Georgia. Boeing produces both the wing and aft fuselage in Seattle, Washington; and Pratt & Whitney builds the F119 engine in East Hartford, Connecticut. An extensive group of secondary and tertiary vendors manufacture subsystems across the country. A discussion of the F-22A vendors is in Chapter Five of this monograph.

Procurement History

The end of the Cold War made the Air Force's decision to continue the stealth fighter program contentious (Walker, 2006).[1] While the Air Force originally planned to purchase 750 aircraft during the Cold War, the per-aircraft procurement costs increased well beyond initial estimates. As a result, the Air Force has continually decreased its planned fleet size. At the time of this analysis, the program of record for the F-22A was 183 aircraft; Congress later provided funding for 4 additional aircraft, bringing the total to 187.

Within the context of this analysis, the F-22A program of record has the following schedule and characteristics:

---

[1] In 1998, the program cost was capped at $37.6 billion by the National Defense Authorization Act for fiscal year (FY) 1998, but the act did not stipulate the total number of aircraft to be procured.

- Production ends with 183 aircraft.
- Last aircraft delivery occurs in 2011.
- Production shutdown activities begin in FY 2009 for some parts of the production line.
- The F-22A sustainment program is based on 183 aircraft.
- The F-22A modernization program is based on 183 aircraft.

Purpose of This Monograph

The F-22A manufacturing base is highly complex and involves multiple long-lead items. In advance of the decision to end production at 187 aircraft, the Air Force asked RAND Project AIR FORCE to evaluate courses of action (COAs) for the F-22A industrial capability when full-rate production is not an option.[2] This monograph presents the results of our analyses. We evaluate three options—or scenarios—and compare our results with continued production at the rate of 20 aircraft per year, the annual rate of the current multiyear procurement (MYP) contract for Lots 7–9 in FYs 2007–2009. Each production option reflects a total aircraft buy of 75. This total was limited because the contractors' labor rate and overhead rate forecasts we used in the analysis of recurring costs extend only to calendar year 2016. The options are defined below.

Option 1. Shutdown

At the conclusion of Lot 9 and delivery of the last aircraft, the production line would close permanently. All tooling, special test equipment, and assembly fixtures would be disposed of or reallocated to other activities, such as current F-22A sustainment or modernization work or other programs.[3] Professional employees and production work-

---

[2] We use COAs and scenarios interchangeably throughout this monograph.

[3] Tools required for a future service life extension or upgrade programs are not included in this option since the Air Force has neither defined such a program nor identified the required tools, test equipment, or technical information for it.

ers would be either reassigned or removed. Production facilities would be reconfigured and reassigned for other work or vacated completely.

Option 2. Shutdown and Restart

Here we posit that the production would shut down and then restart after a two-year gap in procurement funding. We call this option "Shutdown and Restart."[4] While shortening (lengthening) the gap would decrease (increase) the cost of maintaining production capability during the gap and nonrecurring costs during restarted production, Birkler et al. (1993, p. 61), found little correlation between the length of the gap and the recurring production cost at restart.

The Shutdown and Restart option requires significant planning effort for storage and maintenance of most tools, test equipment, technical information, and production facilities. Further, due to the complex nature of the F-22A design and manufacture (especially in such final assembly operations as the integration of specialized electronics and the application of coatings), a core of highly trained engineers and production workers must be retained to make a smooth production restart possible.

Option 3. Warm Production

Under Warm Production, all the means of production would stay in place, albeit used inefficiently, to produce a small number of aircraft during a period of slowdown, until such a point when resources are available to ramp up to the full rate of 20 aircraft per year. In this scenario, five aircraft would be produced per year for a period of three years. This option would require a reduction in force, mothballing of some equipment and tooling, and subsequent rehiring and training of workers.

---

[4] In previous RAND research (Birkler et al., 1993), a "Smart Shutdown" means a shutdown in which actions were taken to facilitate a quick and inexpensive reopening. We use "Shutdown and Restart" to indicate the same concept.

Option 4. Continued Production

Under this option, production continues at the current rate of 20 aircraft per year.

Table 1.1 displays the production profiles for these options.

Our Research Methodology

To estimate costs for each production scenario, we divided costs into two categories: recurring and nonrecurring. Recurring costs are those costs that persist after a production slowdown and restart; these include costs associated with a loss of learning and the effect of each COA on labor rates, overhead costs, and material costs. Nonrecurring expenses are the one-time shutdown and restart costs associated with facility shutdown, rehiring and training of personnel, tool maintenance during shutdown, etc. We estimated these costs based on historical shutdown and restart costs adjusted based on a qualitative comparison between historical programs and that of the F-22A.

We also qualitatively address how F-22A sustainment and modernization contracts as well as the technical data package and contract closeout could be affected by each course of action.

Lastly, we analyzed the large industrial base involved in F-22A fabrication. Vendors were divided into tiers: prime and secondary. The

Table 1.1
Production Quantities, by Scenario

	FY 2010	FY 2011	FY 2012	FY 2013	FY 2014	FY 2015	FY 2016	Total
Option 1	—	—	—	—	—	—	—	0
Option 2	—	—	5	10	20	20	20	75
Option 3	5	5	5	10	20	20	10	75
Option 4	20	20	20	15	—	—	—	75

NOTE: Each production option assumes a total of 75 production aircraft after Lot 9. This assumption was made to keep the Continued Production lot sizes comparable to previous F-22 production lot sizes and the total quantity comparable among all production options. Also, contractor-developed rates and factors did not allow for analysis beyond FY 2016.

secondary vendors were further divided into three categories: high-cost, critical, and randomly selected. This analysis relies heavily on the subjective assessments made by vendor managers at the prime contractors, which may not reflect the views of the vendors themselves. Comparing responses among prime contractors under these circumstances can be difficult. Also, nearly all the information we rely upon is proprietary, and it is difficult to verify its accuracy. Finally, in some instances, the prime contractors provided different levels of detail in their responses and for various reasons data were not always available.

Research Factors

We considered the following assumptions and factors in our analysis:

- All costs are presented in FY 2008 dollars (FY08$), and top-level costs are presented in then-year dollars (TY$).[5]
- All nonproduction-related hours were costed at $185 per hour, an estimate slightly above the fully burdened hourly rate in the military aircraft industry in FY 2008.[6]
- The program of record assumes no additional production after the last aircraft in Lot 9 is delivered, and the sustainment and modernization programs are fully funded with that assumption in mind.
- Future production contracts are annual single-year procurements and not MYP contracts.
- Where possible, we performed sensitivity analysis and provide a range of low, likely, and high cost estimates.
- Operation and support costs to include basing requirements for the additional 75 aircraft were not estimated.
- Government costs were not estimated.

---

[5] FY 2008 dollars are converted into TY dollars through the application of the USAF raw inflation indices for Aircraft & Missile Procurement (3010/20), available from the Deputy Assistant Secretary for Cost and Economics (SAF/FMC).

[6] *Fully burdened* labor rates include direct wages, direct benefits, and apportioned overhead costs. This $185 per hour figure was estimated through regression analysis of wage rates implied in Contractor Cost Data Reports for manufacturing workers in aircraft production from 1996 to 2002.

- The industrial base analysis relied on assessments from the prime contractors using RAND survey instruments and criteria.

Organization of the Monograph

Chapter Two provides an overview of nonrecurring and recurring costs for each option, and then shows our estimate of nonrecurring costs. Chapter Three explains our recurring production cost estimate and summarizes other recurring and nonrecurring estimates. Chapter Four outlines other issues such as sustainment, modernization, and technical data package and contract closeout, and provides a qualitative assessment of how these factors will influence each option. Chapter Five contains an assessment of how the F-22A industrial base is affected if production shuts down for two years. Finally, Chapter Six provides our final conclusions and observations.

There are also three appendixes. Appendix A contains the questionnaires we used to solicit information from the prime contractors about the effects of shutdown and restart on the vendor base. Appendix B presents a short synopsis of lessons from other program shutdowns and restarts documented by the RAND studies by Birkler et al. (1993) and Younossi et al. (2001). Appendix C provides analysis of the original and restart learning curve slopes used in this study.

CHAPTER TWO

Shutdown and Restart Cost Estimates

This chapter describes how we estimated the total cost of procuring F-22A aircraft under each scenario. The first section below gives an overview of the chapter and then recapitulates the scenarios and outlines their requisite shutdown, restart, and production, and support activities. The second section provides estimates for the activities underlying shutdown and restart of the F-22A production line. The third section discusses the procurement cost model and then focuses on how loss-of-learning at production restart and changes in procurement quantity affect unit cost. The fourth section presents estimates for the costs of flyaway and below-the-line elements (munitions, support, sustainment, spares, logistics, retrofits, and other government costs [OGC]) in the F-22A procurement budget. The final section compares the total costs of pursuing each COA at the flyaway and total procurement cost levels.

The activities and costs described in this chapter encompass the entire production program: airframe (including structures, systems, and coatings), avionics and other electronics, propulsion, and their integration. However, the description of Shutdown and Restart activities focuses mainly on airframe, avionics, and overall integration. This is because Pratt & Whitney operates differently from other contractors: Its use of its own facilities as well as modular processes and purchasing strategies across all its engine lines implies that shutting down and reconstituting its production capability is less predictable but also requires less government management.

Types of Cost and Profit

This study used a variety of methods to estimate production and non-production activities, as well as several top-level cost measures, as seen in Table 2.1. Gap-related costs include all activities specific to hiatus (shutdown with intent to restart), restart, and termination (end-of-production shutdown), with a fee of 10 percent. Target price curve (TPC) and propulsion costs include the costs of airframe, avionics, and engines for the F-22A, including a profit of 13.52 percent. Flyaway costs include not only TPC and propulsion costs, but also an additional set of activities grouped as other flyaway elements. Total cost includes all gap-related costs, flyaway cost, as well as other below-the-line elements.

The cost categories in Table 2.1 appear in the order in which they are addressed in this and the following chapter.

Table 2.1
Cost Categories, by Type of Activity

	Gap-Related Costs	TPC and Propulsion Costs	Flyaway Costs	Total Cost
Termination	X			X
Hiatus	X			X
Restart	X			X
Production				
Airframe		X	X	X
Avionics		X	X	X
Engine		X	X	X
Tail-up		X	X	X
Other flyaway elements			X	X
Below-the-line elements				X

Activities Are Tied to Production Profiles

We took a straightforward approach to estimating the cost of each scenario. First we identified the scope and timing of all termination, hiatus, restart, production, and support activities that would need to be done. Then we estimated the cost of each activity using one or more methods, including large-scale cost modeling, extrapolation from recent historical data, historical analogy, and subjective judgment, when required. We tied all activities to the production time line associated with each scenario. Table 2.2 presents the production time lines for each of the four scenarios we analyze—Shutdown, Shutdown and Restart, Warm Production, and Continued Production—with their annual production quantities.

Types of Activities

This time-line view, while useful for budgeting, is less useful for estimating. We relied on an activity-based view, presented in Table 2.3, that places the activities necessary to procure the aircraft into four large groups and lists by scenario the fiscal years in which they must be performed. The four groups of activities are hiatus, restart, production, and termination.

Table 2.2
Annual Production Quantities, by Scenario

Scenario	FY 2009	FY 2010	FY 2011	FY 2012	FY 2013	FY 2014	FY 2015	FY 2016
Shutdown	20	—	—	—	—	—	—	—
Shutdown and Restart	20	—	—	5	10	20	20	20
Warm Production	20	5	5	5	10	20	20	10
Continued Production	20	20	20	20	15	—	—	—

Table 2.3
F-22A Activities, by Scenario and Fiscal Year

Scenario	Hiatus	Restart	Production	Termination
Shutdown	—	—	—	2009–2011
Shutdown and Restart	2009–2011	2010–2012	2012–2016	2016–2018
Warm Production	—	2013	2010–2016	2016–2018
Continued Production	—	—	2010–2013	2013–2015

For the most part, these activities inside these four groups are self-explanatory. Termination includes all the labor and planning involved in breaking down and removing tooling, recording data, managing excess inventory, and cleaning up facilities. Hiatus involves the same type of activities as termination, but with an emphasis on the preservation of production capability; hence, tools are stored and maintained, instead of scrapped. Restart activities are in large part the inverse of hiatus activities: rehiring and retraining personnel, reorganization of facilities, reassembling tooling—in short, facilitating the production process, and "ramping-up" production quantities. The costs of termination, hiatus, and restart activities are estimated in various ways.

Production includes every cost of procuring the F-22A: from material and equipment, to hourly labor and salaried engineering workforce, to sustainment and support activities. We modeled the core of these costs, in order to estimate the impact of a production hiatus and changes in production rate on unit cost. In addition, we extrapolated from historical expenditures to estimate all other support and sustainment costs in the production budget.

Note that under all scenarios production terminates at some point: this ranges from FY 2009 in the Shutdown scenario, to FY 2013 under the Continued Production scenario, to FY 2016 for both the Shutdown and Restart and Warm Production scenarios. The content of this end-of-production shutdown does not vary across scenarios. The content of the rest of the activities varies by scenario:

- For Shutdown, production ends in FY 2009, and production termination, which maintains no capability of restart, is performed; there is no production hiatus or restart, or further production. Under this scenario, the Air Force does not procure an additional 75 aircraft.
- For Shutdown and Restart, the production line is placed into hiatus in FY 2009, meaning that actions (like tool storage) are taken to ensure an efficient and timely return to production in FY 2012; however, restart costs occur almost immediately upon hiatus. Production continues until FY 2016, when it terminates.
- For Warm Production, production continues at low rates starting in FY 2010; starting in FY 2013, full-production restart costs are incurred with quantity ramp-up, because contractors must hire and train more workers. Production continues until FY 2016, when it terminates.
- For Continued Production, there is no hiatus and no restart because the production line operates unchanged until three-quarters of the way through FY 2013, when production is terminated.[1]

Shutting Down and Restarting the Production Line

This section contains a detailed description of the activities that would permanently end F-22A production as well as those that may re-enable F-22A production after a production gap or persistent reduction in rate. First, we estimated termination costs, because they constitute the only activity of the shutdown scenario and are a required activity of the others. Second, we examined the similarities and differences between production termination and placing production on hiatus. Third, we examined the activities required for restart, looked at their links to activities in hiatus, and estimated those costs.

---

[1] The production of 15 units is assumed to occur at the same rate as the production of 20 units, but for three-quarters of the time.

Termination

Overview. Under every COA, the F-22A production line will eventually close, with complete disposition of all tools, test equipment, facilities, personnel, data, and inventory—in short, without maintaining the ability to restart. This end-of-production termination will always occur during and at the end of the last production run for each scenario: FY 2009 for Shutdown, FY 2016 for Shutdown and Restart and Warm Production, and FY 2013 for Continued Production.

In several historical cases, termination has been divided into two phases. In the first phase, the government requests an initial evaluation by the prime contractors, providing a list of activities and a rough estimate of associated costs. In the second phase, the government requests that the prime contractors make a firm cost quotation derived from a thorough analysis of the shutdown process, including preparing detailed plans and procedures for finishing up production; accounting for all government-owned tools, test equipment, and remaining inventory; and ensuring that all contract deliverables have been made. Once executed, this phase includes the removal and transfer or disposal of all tooling, test equipment, and inventory; the clearing of facilities; and the validating and transfer of plans and program management data.

We have included both the planning and execution phases in our termination estimate, but RAND's estimate itself must be considered less detailed than those that would be prepared by dedicated personnel with the technical expertise at a prime contractor and system program office. The various activities comprising production termination were aggregated into four categories: tooling, facilities, inventory, and planning and administration. In addition, the dollar values presented here are for the near future—2008 to 2011. Yet they should not be interpreted as precise and are not inflation-adjusted or time-phased. The profit percentage applied here is 10 percent, which we think is reasonable for termination efforts, although the default bargaining position for negotiations is usually 15 percent.

Tooling and Special Test Equipment. Tooling activities include accounting for all government-owned and prime contractor–owned tools and special test equipment at all prime contractor locations and suppliers by tagging and entering them into an information management system, and either boxing and shipping them where the government indicates or having an outside company disassemble and scrap them. Ensuring the integrity of the termination process requires a considerable amount of record keeping and accounting; immediately scrapping all tooling is not a feasible option.

A detailed bottom-up estimate requires that the type of tasks needed to be performed, and the number of hours needed to perform them, be derived from a detailed manpower assessment. We used the top-level approach described below.

As can be seen in Table 2.4, we first estimated the number and size of the tools, then varied the hours necessary to inspect, tag, box, and transfer them. The number of tools comes from a precise number provided by Lockheed Martin for its own F-22 and supplier operations, combined with our own estimate of the number of tools managed by Boeing and Pratt & Whitney and their suppliers.[2] A more precise estimate would use up-to-date and validated numbers from the contractor's tooling information management system.

A possible number of hours for each task was provided by Lockheed Martin in a related estimate of F-22A shutdown costs. We have modified the content of that scenario to fit our definition of termination and have modified the number of hours to generate low and high estimates. All tasks incorporate supervisory and hand-labor efforts. Note that some large tools are exceedingly cumbersome and require a considerable disassembly effort. However, we believe that many small tools will be processed in batches; therefore, the number of hours actually needed to process small tools will be less than estimated.

Under our termination estimate, nearly all F-22A-unique production tooling would be scrapped.[3] The process envisioned is one in which contractor personnel move tooling outside of the facility, and a third-party scrap operator removes it from the premises. The scrap-

---

[2] Estimates of number of tools provided to the Air Force by contractors are substantially higher than those we used in our cost estimates. Classified tools were not distinctly accounted for and may require an additional premium to handle.

[3] This tooling does not include the tools and test equipment needed for future depot maintenance, which are already in place. It also does not include tooling used to produce parts for systems other than F-22A.

Table 2.4
Cost of Disposing of Tooling Under Termination

	Low			Likely			High		
Type of tool	Large	Small	Total	Large	Small	Total	Large	Small	Total
Total number of tools	2,736	5,900	8,636	2,736	5,900	8,636	2,736	5,900	8,636[a]
Record retention hours	4	4	—	5	5	—	6	6	—
Planning hours	4	3	—	6	4	—	8	6	—
Breakdown hours	12.5	0.5	—	14.5	0.5	—	17	0.5	—
Scrap preparation hours	2	1	—	3.5	1.5	—	4	2	—
Total hours per tool	23	9	—	29	11	—	35	15	—
Total hours (thousands)	61.6	50.2	111.7	79.3	64.9	144.2	95.8	85.6	181.3
Cost per hour ($)	—	—	185	—	—	185	—	—	185
Total cost (FY08 $million)	—	—	20.6	—	—	26.7	—	—	33.5
Total cost plus profit (FY08 $million)	—	—	23	—	—	29	—	—	37

NOTE: Totals have been rounded.
[a] Estimates provided after completion of this analysis were substantially higher.

ping itself was considered a zero-cost, zero-profit activity, and was not costed separately.

Facilities. Facilities activities include the complete clearing out of all F-22 production buildings under conditions yet to be specified in agreements between contractors and the government.[4] We made our cost estimates on a cost-per-square-foot basis and varied the cost per square foot.[5] As can be seen in Table 2.5, using square-footage data obtained on visits to prime contractor sites, we estimated the current

[4] At the prime contractor level, manufacturing and integration are performed at government-owned, contractor-operated sites in Marietta and Fort Worth (Lockheed Martin), and elsewhere. Seattle (Boeing) and Middletown and East Hartford (Pratt & Whitney) own and operate their own facilities.

[5] The cost per square foot to clear a facility is intimately tied to the density of wiring, cabling, hand-tooling, and support equipment that is built into the structure of the facility.

Table 2.5
Disposition Costs of Facilities (FY08$)

	Low	Likely	High
Cost per square foot	$4.5	$6	$7.5
F-22-dedicated square feet	2,145,000	2,145,000	2,145,000
Shared square feet	643,500	643,500	643,500
Total square feet	2,788,500	2,788,500	2,788,500
Total cost ($million)	12.5	16.7	20.9
Total cost + profit ($million)	14	18	23

NOTE: Totals have been rounded.

dedicated F-22 nonproduction and production floor space at Boeing and Lockheed Martin to be 2.145 million square feet.[6]

This area estimate was increased by 30 percent to account for floor space mixed between F-22 and other production runs, yielding 2.789 million square feet. Low, likely, and high estimates of $4.5, $6.0, and $7.5 per square foot, yield, after applying 10 percent profit, estimates of $13.8 million, $18.4 million, and $23.0 million.[7] We believe our estimate of facilities-clearing costs is reasonable.

Inventory. Inventory activities can be considered as being of two types: management and identification of all inventory leading up to termination using existing personnel and systems, and packaging and preparing the remaining inventory for shipment.

To estimate likely inventory disposition costs, we used the methodology of escalating the excess inventory at the end of the F-14 run. The F-14 program contracted for a total cost of roughly $8 million in "excess material" at the end of production, which occurred in FY 1990. Escalating these costs using the USAF procurement index yielded $11.5 million in FY 2008 dollars.

---

[6] The Air Force estimates for floor space required to store tooling are somewhat lower than our estimates.

[7] As a reasonableness check on these figures, we note that at $185 an hour, $6 per square foot implies $185 / $6 = 30.8 square feet per hour. A 100-square-foot cubicle would take one person 100 / 30.8 = 3.25 hours to clear.

Our understanding of this task presumes that most packaging and preparation is unwarranted; our justification is that since the government has no current plans to use that material, it should not pay to pack and prepare (and eventually store) material it cannot manage or use. Hence, in our estimate, excess inventory would be either trashed or scrapped at no cost. The F-14 program, however, paid to pack and prepare its inventory to be shipped, so we discount the $11.5 million estimate by an arbitrarily chosen 10 percent, yielding a final likely cost of $10.3 million. Low and high estimates were created by adjusting remaining inventory by $5 million down and up respectively.

Planning and Administration. Planning and administration activities are fourfold: required restart planning analysis and documentation; executive and engineering administration of termination activities; aggregation and transfer to the government of program management data and lessons learned; and continued consulting analyses of the F-22 program shutdown. Restart planning documentation includes initial and follow-on contract creation, as well as the creation and operation of systems and reports to demonstrate to government program management that the termination is being conducted according to contract and in an efficient and timely manner. Executive and engineering administration of termination activities pays for a small contractor team (one to two people at each prime contractor), which has oversight and knowledge of all termination activities and which responds directly to government program office queries. Program data and lessons learned do not include detailed design data;[8] instead, they include items such as vehicle analysis data (e.g., aerodynamic and thermodynamic performance), problem and report databases, and flight test data. However, data and lessons learned of specific importance to the government have not yet been identified. Outside consulting includes studies and support from research organizations and consultants.

---

[8] Maintenance of a technical data package including all design specifications has been contracted for in the sustainment contract. Essentially, the data and their upkeep are leased by the government on an annual basis and need not be purchased.

The need for such supporting activities was intimated in our discussions with F-22A contractor and program office personnel, but requirements for such activities have not yet even been outlined. Accordingly, our estimating methodology is highly subjective, utilizing analogy to other activities. The $9.0 million for restart planning documentation is derived from several historical contract efforts; $1.0 million per year for executive and engineering management is consistent with the number of oversight managers planned; $7.0 million for maintaining program management data is an arbitrary allocation; $2.2 million for outside analysis is consistent with several major outside analyses. The likely estimate for these tasks was generated on a fiscal year profile, as shown in Table 2.6. This yields a likely estimate of $21 million; low and high estimates of $17 million and $27 million, respectively, were created by subtracting and adding 25 percent from the likely estimate.

Summary of Termination Costs.[9] For the Shutdown scenario, a low, likely, and high cost of each category within termination costs are listed in Table 2.7. These estimates represent our reasoned judgment of the cost of the different levels of effort needed to complete each activity. Likely costs ($79 million) were estimated first, with low ($59 million) and high ($102 million) variations resulting primarily from parametric changes in labor hours needed to perform the same tasks. Low and

Table 2.6
Administrative and Planning Cost During Termination (FY08 $million)

	2010	2011	2012	Likely Total
Restart planning documentation	3.0	3.0	3.0	9.0
Executive and engineering management	1.0	1.0	1.0	3.0
Maintain program management data	4.0	2.0	1.0	7.0
Outside analysis	1.3	0.6	0.3	2.2
Total[a]				21.0

[a] Total has been rounded.

[9] Our cost estimate is based on storing 8,636 tools, whereas the new estimate of the number of tools from the Air Force is substantially higher.

Table 2.7
Low, Likely, and High Estimates of Termination Costs for Shutdown Scenario (FY08 $million)

	Termination Cost for Shutdown Scenario		
	Low	Likely	High
Tooling	22.7	29.4	36.9
Facilities	13.8	18.4	23.0
Inventory	5.3	10.3	15.3
Planning and administration	17.0	21.2	26.5
Total	59	79	102

NOTE: Totals have been rounded.

high values are not risk estimates and were not generated from probability distributions. The wide range ($43 million) from low to high indicates the paucity of relevant and detailed historical information. These ranges will begin to narrow when the F-22A shutdown plan has been designed and activities within it can be described in detail.

As seen in Table 2.8, the likely then-year costs for the Shutdown and Restart, Warm, and Continued scenarios differ only slightly, since the only difference between them is a shift in time. While termination activities are the same for all scenarios, the year varies: FY 2009 for Shutdown, FY 2013 for Continued Production, and FY 2016 for Warm Production and Shutdown and Restart. Inflation turns an

Table 2.8
Likely Termination Costs, by Scenario (TY $million)

	Shutdown FY 2009	Shutdown and Restart FY 2016	Warm Production FY 2016	Continued Production FY 2013
Tooling	31.1	35.8	35.8	33.7
Facilities	19.1	22.0	22.0	20.7
Inventory	10.7	12.3	12.3	11.6
Planning and administration	22.4	25.7	25.7	24.3
Total	83	96	96	90

NOTE: Totals have been rounded.

$83 million then-year cost in the Shutdown scenario in 2009 to a $90 million cost for Continued Production in 2013, to a $96 million cost for Shutdown and Restart and Warm Production in 2016.

Hiatus

Overview. When it terminates production, the Air Force will transition some dual-use production equipment to sustainment, and it may choose to preserve lessons learned. However, it will not preserve the detailed information, tools, equipment, and personnel necessary to reproduce the F-22A easily.

The essential difference between production hiatus and production termination is *preservation*: the maintenance of engineering data and reports, material and equipment sourcing records, drawings, financial data, tooling, facilities, and key personnel essential to build the same configuration aircraft after restart as before shutdown.[10] Table 2.9 shows our range of estimates for the cost of the activities related to hiatus.

Since no specific Shutdown and Restart plan or array of plans has been developed, we chose a hiatus plan with as much overlap as possible with termination. For facilities, inventory, and planning, we

Table 2.9
Costs of Shutdown and Restart Scenario (FY08 $million)

	Production Hiatus in the Shutdown and Restart Scenario		
	Low	**Likely**	**High**
Tooling	29.0	54.1	71.1
Facilities	13.8	18.4	23.0
Inventory	5.3	10.3	15.3
Planning and administration	17.0	21.2	26.5
Total	65	104	136

NOTE: Totals have been rounded.

[10] The core activities that need to be performed were detailed in a previous RAND analysis (Birkler et al., 1993, pp. 21–29).

assume that the range of estimates provided for termination are wide enough to encompass all similar tasks while preparing for hiatus. Again, the low, likely, and high costs here represent high-level estimates of a "drape-in-place" shutdown with intent to restart. In reality, some facilities will be kept "dark" until restart, other facilities will be cleared out entirely and transferred to other locations permanently, and still other facilities may be transitioned to uses other than production, depending on what is determined to be the most profitable alternative for contractors when hiatus begins.

Discussions with prime contractors led us to believe that it is not critical to maintain the layout of the existing production line, but it *is* critical to preserve the knowledge of how to build the current configuration efficiently while actively managing and maintaining the supplier base. Although draping-in-place of all tools and test equipment in current facility space provides the appearance of maintaining a production capability, it does not maintain production knowledge, a skilled workforce, the information retained in engineering, process, financial systems, lessons learned, or any real form of reconstitution capability.[11] Under termination, government-owned facilities are cleared completely and transferred back to the government, but under hiatus, our high-level estimates of the costs of facilities are for clearing and securing employee workstations and areas, locking down facilities, and keeping them air-conditioned and clean. We assume that inventory costs and planning costs would be no different in termination and hiatus, because inventory would be disposed of in the same manner and similar planning issues are involved.

Tooling. Compared with production termination, a production hiatus includes costs for packing, inspecting, and storing tooling for 18 to 24 months, instead of scrapping them.[12] The costs would be higher for a longer hiatus. As can be seen in Table 2.10, compared with our

---

[11] Later information provided to RAND indicates that there are several extraordinarily large F-22A-unique tools that will almost certainly remain draped in place. This fact limits, but does not prevent, the later reorganization of the production line.

[12] Under the current domestic acquisition environment and the prohibition of F-22A foreign military sales, it is unlikely that the prime contractors would choose to preserve future production capability with their own funds.

Table 2.10
Cost of Disposing of Tooling Under Hiatus (FY08$)

	Low			Likely			High		
Type of tool	Large	Small	Total	Large	Small	Total	Large	Small	Total
Number of tools	2,736	5,900	8,636	2,736	5,900	8,636	2,736	5,900	8,636
Records retention hours	4	4		5	5		6	6	
Planning hours	4	3		6	4		8	6	
Breakdown hours	12.5	0.5		14.5	0.5		17	0.5	
Storage preparation hours	4	2		16	8		20	10	
Total hours per tool	25	10		42	18		51	23	
Thousands of hours	67.0	56.1	123.1	113.5	103.2	216.8	139.5	132.7	272.3
Cost per hour ($)	—	—	185	—	-	185	—	—	185
Tooling preparation cost ($million)	—	—	22.8	—	—	40.1	—	—	50.4
Tooling preparation cost + profit ($million)	—	—	25.0			44.1			55.4

NOTE: Totals have been rounded.

tooling estimate for termination, our tooling estimate for hiatus excludes the labor required for scrapping (dragging the tools outside the buildings), and includes far more extensive labor to prepare for storage, yielding a likely tooling preparation estimate of $44 million and a range of $25 million to $55 million.

The cost of storage was estimated on a cost-per-square-foot basis, as presented in Table 2.11. Using data provided by the contractors, we estimated that all tooling could be stored in roughly 750,000 square feet of space for roughly 2.5 years. However, due to the relatively short nature of the full hiatus in the Shutdown and Restart scenario and a longer period of low production, it is likely that a smaller storage footprint will be needed for a longer period of time. We have estimated that this equates to roughly 1.75–2.75 years of full storage space of 750,000 square feet. However, the cost of this storage is also uncertain and could range from a low of $3 per square foot per year to $5.3–$7.6, including profit. Hence, our likely estimate for tool storage during hiatus is 2.5 × $5.3 × 750,000 = $10 million, with a range of $4 million to $16 million. The likely total tooling estimate, the sum of the preparation estimate and storage estimate, is $54 million, with a range of $29 million to $71 million.

Even after including the premium for managing tooling, the activities listed in Table 2.11 are not the only requirements for a production hiatus. Only those activities charged directly to F-22A contracts are accounted for in Table 2.11. Some other requirements, such as technical drawings and laboratory facilities, will be paid for by the Air Force under the F-22A sustainment contract regardless of which COA the Air Force chooses.[13] Other costs, such as the reduction in force required at the completion of Lot 9 production, could be considerable, but are charged indirectly to overhead, which is charged to all programs at contractor facilities.

---

[13] Fully funded sustainment and modernization contracts are the baseline from which the shutdown with intent to restart is estimated.

Table 2.11
Cost of Tooling Storage Under Hiatus (FY08$)

	Low	Likely	High
Thousands of square feet	750	750	750
Years	1.75	2.5	2.75
Storage cost per square foot ($)	3	5.3	7.6
Tooling storage cost + profit ($million)	4	10	16
Tooling preparation cost + profit ($million)	25.0	44.1	55.4
Total tooling cost + profit ($million)	29	54	71

NOTE: The production tooling quantity estimate that the Air Force obtained from industry after the completion of RAND's research is higher than that provided to RAND earlier.

Restart

Overview. Two types of activities are lumped into restart: (1) those that are needed to maintain capabilities or prevent problems during hiatus that are essential to timely and cost-effective restart, and (2) those required to start up production. Table 2.12 lists those activities and associated costs.

The specific restart activities that must be performed depend on the path chosen at the start of production hiatus; however, even a detailed Shutdown and Restart plan could lead to multiple restart options. Nevertheless, the broad array of activities will in many ways mirror hiatus: personnel who have been let go must be rehired; tooling that was stored must be brought back and calibrated; facilities that

Table 2.12
Costs of Restart Activities (FY08 $million)

Activity	Low	Likely	High
Personnel rehired	30	56	71
Tooling reorganization	25	44	55
Facilities setup	14	18	23
Planning and administrative tasks	9	18	27
Personnel maintained	32	65	97
DMS[a]	—	—	—
Requalification	40	129	281
Total	150	330	554

NOTE: DMS = diminishing manufacturing sources.

[a] DMS is ordinarily a "flyaway" cost, not a shutdown or restart cost. It is included in the production cost estimate in the next chapter.

have been cleared must be set up, or if original facilities have been dedicated to other uses, new space must be set up; and planning and administrative tasks continue, because those who once oversaw the last units of production and management of the hiatus now manage the reconstitution of the assembly lines, the supply chains, and the testing facilities.

Personnel. While the costs of labor force reductions at prime contractors are indirect costs that will affect the overhead rates of all programs and have been captured in the baseline program, the costs of rehiring, training, and getting clearances for workers are a mix of direct and indirect costs that are not accounted for elsewhere. Our estimate of these personnel costs can be seen in Table 2.13. We used our F-22A production cost model (described below) to generate the hours required to produce aircraft and then translated those hours into the number of salaried and hourly workers who must be rehired to complete the work. The number of workers who must be rehired is the same for the low, likely, and high estimates.[14] Based on our discus-

[14] We present three cases (low, likely, and high) to reflect differences in retained learning, which affects primarily how early in the restart process workers must be rehired, not the total number that eventually need to be rehired.

Table 2.13
Costs of Rehiring Personnel for Restart (FY08$)

	Salaried Workers	Training Days	Hourly Cost ($)	Total Cost ($million)
Low	279	10	185	4.1
Likely	279	10	185	4.1
High	279	15	185	6.2
	Hourly Workers	**Training Days**	**Hourly Cost ($)**	**Total Cost ($million)**
Low	1,749	10	185	25.9
Likely	1,749	20	185	51.8
High	1,749	25	185	64.7
	Total Cost ($million)			
Low				30
Likely				56
High				71

NOTE: Totals have been rounded.

sion with the contractors, we posit that between two and five weeks' worth of pay for each new hire would cover administrative, training, and clearance costs, providing a range of estimates for the total cost of rehiring.[15] Discussions with contractors led us to believe there was far greater uncertainty about the range of training days for hourly workers than for salaried workers.

The number of personnel to be rehired depends on many factors, including whether any personnel were specifically targeted to maintain key production expertise. The "personnel maintained" category in Table 2.12 has 50, 100, or 150 full-time workers dedicated to the maintenance of the F-22A production program over the two-year shutdown. Given a relatively short gap and the need to rehire person-

[15] Clearance costs are twofold: internal management and idle time between hiring and clearance approval. The former are overhead costs, and the latter are either zero (if a worker already holds a clearance) or up to roughly eight weeks (if a worker must wait for a clearance). Our estimate includes a very small number of rehires who will have extensive idle time.

nel upon restart, it may be cost-effective to retain knowledge of the production program by maintaining key production personnel with the specialized skills necessary for F-22A manufacturing. Discussions with contractor and program office personnel revealed that maintaining expertise within the F-22A program is critical for a timely and efficient restart; however, the number of personnel needed would require a detailed analysis. Their jobs during the shutdown would be to keep the program ready for restart, maintaining critical information systems as well as institutional experience and lessons learned.

Tooling, Facilities, and Planning. Tooling, facilities, and planning and administrative costs are assumed to mirror those of hiatus: the same activities in reverse with the same types of costs, escalated in cost to different fiscal years, with more variation. While this may seem to be an inadequate assumption and an overly broad estimate of restart costs, they must be understood as a cost estimate at a very high-level vision of restart; there is simply no way to determine a precise cost for a set of activities that have not been identified.

For restart tooling cost, we used the hiatus cost of tooling (without storage) listed in Table 2.11. For restart facilities cost, we used the hiatus facilities cost.

Restart planning and administrative costs are outlined in Table 2.14. It includes a significant amount of funding to reestablish and validate reporting databases as well as funding for a limited number of contractor personnel to oversee restart operations. Other funding includes support to the F-22 program office as well as specific funding to document lessons learned from F-22 restart.

Diminishing Manufacturing Sources and Requalification. The other two types of activities performed during the gap both relate to a supplier's inability to provide the same parts at production restart as at the start of hiatus. We call these activities DMS and requalification. *DMS parts* are those that can no longer be bought in the marketplace because of a number of factors, including technological obsolescence. DMS cost is a regular part of all

Table 2.14
Restart Administrative Costs (FY08 $million)

	2013	2014	2015	All Years
Database and reporting validation	6.0	3.5	1.0	10.5
Executive and engineering management	1.5	1.0	1.0	3.5
Document restart lessons learned	0.5	0.5	1.0	2.0
Outside analysis	1.0	0.6	0.3	1.9
Total cost	9	6	3	18

NOTE: Totals have been rounded.

normal production programs, requiring active management and part buyout for all expected future aircraft or part redesign. DMS cost has varied considerably from year to year on the F-22A program; it is inherently unpredictable, and annual budgets have included funds for unforeseeable DMS "pop-up" costs. To ensure the timeliness of a planned restart, ordinary DMS must continue as if production were under way.

Yet DMS is ordinarily a flyaway cost, not a shutdown or restart cost, and we include it in the production cost estimate below. This is the reason DMS cost is footnoted in Table 2.12; the costs will have to be paid during hiatus and are required for restart, but they are actually production costs and will be accounted for in the production cost estimate. This implies that under the Shutdown and Restart scenario, the F-22A program will experience a unique cost accounting issue: having flyaway costs in fiscal years (FY 2010 and FY 2011) when no units are being produced.[16]

The final restart cost is for requalification. *Requalification* activities are those taken by the government and prime contractors to mitigate the need for supplier requalification during restart. They can be thought of as "extraordinary" DMS issues that result from the lack of

[16] Historical analysis of the F-22A's production history since Lot 1 indicates that from $43 million to $175 million is required on an annual basis to handle DMS, with an average of $95 million; this is incorporated in the production cost estimate in the next chapter.

continuous production, over and above the DMS ordinarily faced by production programs.

Estimating requalification costs is not a direct process. We did not extensively study the actions that would be necessary to ease the requalification process. Instead, we asked program office officials to identify those firms at highest risk of difficulties with requalification and the problems they would face. Those firms, 11 in number, and their deliverables to the F-22 program were analyzed for product cost and complementarities to other firm products. To derive a cost for requalification, we chose a subjective probability of the firm facing the identified problem and multiplied it by the historical cost of overcoming a similar difficulty during production. An aggregation of these judgments can be seen in Table 2.15. The likely cost of restart requalification can be thought of as roughly having to overcome one small requalification issue, one of medium difficulty, and one of very great difficulty.

In some cases, the cost of requalification potentially exceeds the continued production cost of 75 additional aircraft, suggesting that the least-costly and easiest-to-manage method of ensuring parts availability at reasonable costs is advanced procurement of those parts during shutdown. However, this requalification procurement is not permitted under current authorization, and Congress must grant the F-22A program office a special exemption to carry out these purchases. We have not produced a cost estimate under the assumption that these parts will be procured in advance, but for F-22 production to be restarted cost-

Table 2.15
Aggregate Requalification Cost of Restart

Number of Firms	Estimated Cost of Mitigation (FY08 $million)	Low	Likely	High
5	4	10%	20%	50%
1	12	5%	20%	50%
1	50	10%	20%	50%
4	80	10%	35%	75%
Total cost (FY08 $million)		40	129	281

effectively, the program has little choice but to request that Congress authorize and appropriate funds to manage high-risk supplies during a production hiatus.

Warm Production. For the Warm Production scenario, hourly production personnel are let go starting in FY 2010, because the number of aircraft to be produced shrinks from 20 to 5. These are largely indirect costs to all programs. However, these production workers must be rehired starting in FY 2013 as production ramps up. Based on our production cost model (in Table 2.16), we estimate that 858 production workers will have to be rehired.

At $185 per hour for our likely estimate, we posit that 25 days of eight hours each will cover the total training, administration, and clearance cost, yielding a rehiring cost of $32 million.[17]

Table 2.16
Cost of Rehiring Production Personnel

	2013	2014	2015	Total
Number of workers	70	343	445	858
Labor hours	200	200	200	200
Cost per labor hour (FY08$)	185	185	185	185
Total cost (FY08 $million)	2.6	12.7	16.5	32

Summary

The results generated in the previous sections provide a complete picture of the costs of the nonproduction-related activities required to pursue each course of action. These are summarized in Table 2.17.[18] This table should be read row by row. The Shutdown scenario is presented in the first row; only termination applies, and it is

[17] (858 workers) × ($185/hour) × (8 hours/day) × (25 days/worker) = $31.7 million. This is based on discussions with the contractors.

[18] Hiatus costs are from Table 2.9, restart costs are from Table 2.12 and the Warm Production subsection, and termination costs are from Table 2.8.

estimated to cost $79 million. The second row presents Shutdown and Restart; hiatus applies ($104 million), as well as restart ($330 million) and termination ($79 million), for a total of $513 million. Warm Production is in the third row; restart ($32 million) and termination ($79 million) total $111 million. Continued Production is in the fourth row; only termination applies, and it costs $79 million.

Table 2.17
Costs of Shutdown and Restart Activities, by Scenario (FY08 $million)

Scenario	Hiatus	Restart	Termination	Total
Shutdown	—	—	79	79
Shutdown and Restart	104	330	79	513
Warm Production	—	32	79	111
Continued Production	—	—	79	79

CHAPTER THREE

Production Cost Estimate

This chapter shows how we estimated the cost of procuring 75 additional aircraft after Lot 9 under the Shutdown and Restart, Warm, and Continued Production scenarios. For the Shutdown and Restart scenario, we developed a range of estimates—optimistic, average, and pessimistic—depending on how much learning is retained. We estimated the cost using a recurring production cost model developed for a previous RAND F-22A study, modified to take into account the effects of Shutdown and Restart.[1]

Airframe Contractors Target Price Curve and Propulsion Costs

This model estimates costs for TPC and propulsion, leaving other recurring and nonrecurring costs to be estimated independently.[2] Table 3.1 lists the methods used to generate the estimates of total cost.

[1] The details of the original model can be found in Younossi et al., 2007, pp. 19–34.

[2] The TPC includes all costs associated with production of aircraft except the engine and sustaining labor that cannot be uniquely identified with particular aircraft.

Table 3.1
Costs of Restart Activities

Elements of Total Cost	Method
TPC	Labor and material model
Propulsion	Material model
Other flyaway cost elements	Extrapolation from F-22 historical costs
Other below-the-line cost elements	Extrapolation from F-22 historical costs

Overview of TPC and Propulsion Model

Here we review the original model and show how we extended it to estimate the effect of restarting the production line.[3] Unless otherwise stated, the model used in the current analysis is identical in assumptions and structure to the original model.

The overall purpose of both models is to calculate the cost of aircraft lots by fiscal year. The cost of each lot is the sum of labor and material costs, by major vendor and site.[4] Labor cost is a calculation of labor hours times contractors' labor rates. Each prime contractor provided us with a best estimate of future labor rates in response to the specific production scenarios analyzed in this document. Our main concerns are estimating labor hours and material dollars. In the original TPC model and in the current model before a production hiatus, labor hours (and material dollars) are estimated by means of the following cost-improvement curve:

$$C_n = C_1 \times n^{\ln(b)/\ln(2)} \times r^{\ln(c)/\ln(2)} \tag{3.1}$$

where

C_n is the average hours for units in the lot
C_1 is hours for the first unit
n is the lot midpoint unit number

[3] The previous version of this model was generated in a program called Analytica; for the current effort, it was extended in Analytica, and then replicated and validated in Microsoft Excel.

[4] We will describe the methodology used to estimate labor cost. The methodology used to estimate material cost is similar, though not identical, to estimation of labor hours.

b is the unit cost improvement slope
r is the number of units procured in the lot
c is the rate slope.

We applied Equation 3.1 to both labor hours and material dollars. The labor hour and material dollar information comes from historical, validated, completed lot information for Lots 1 through 4, an estimate of completion data for Lot 5, and negotiated contract information for Lot 6. Labor hour data were aggregated into touch (manufacturing) and engineering, for Lockheed Martin and Boeing production sites (Fort Worth, Marietta, Palmdale, and Seattle).

Forecasting future values requires estimating the constants C_1, b, and c in Equation 3.1. Because n, the lot midpoint quantity, and r, the lot production rate, are highly correlated in the F-22 production history, estimating b and c simultaneously would generate spurious results. However, a previous analysis of other aircraft programs estimated Equation 3.1 for other historical programs whose n and r were not highly correlated, resulting in a c for airframes of 0.89 (Arena et al., 2008).[5] We then estimated C_1 and b after plugging in c from the other analysis.

Effects of Changing Annual Procurement Rate

The effects on hours, and costs, of varying production rate are directly estimated from the $r^{\ln(c)/\ln(2)}$ component of Equation 3.1. When 20 aircraft units are procured annually, the rate component measures $0.60 = 20^{\ln(0.89)/\ln(2)}$. But when 10 units are procured annually, the rate component increases 12 percent to $0.68 = 10^{\ln(0.89)/\ln(2)}$; when 5 units are procured annually, the rate component increases another 12 percent to $0.76 = 5^{\ln(0.89)/\ln(2)}$.

A similar analysis was done for engines; however, at 0.97, the c for engines is far higher than the c for airframes. When 40 engines are procured annually, the engine rate component measures $0.85 = 40^{\ln(0.97)/\ln(2)}$. But when 20 engines are procured annually, the engine rate component measures $0.88 = 20^{\ln(0.97)/\ln(2)}$. Halving the

---

[5] Previous RAND analysis (Younossi et al., 2007) estimated c for engines at 0.97.

number of engines procured annually increases the engine rate component by 3 percent.

Selecting the Portion of the Original Curve

To estimate a cost improvement curve and forecast future lot costs, the RAND analysis of F-22A multiyear procurement used cost data for Lots 5 and 6, instead of costs for Lots 1 through 6, or just Lot 6. We argued that using data from just Lots 5 and 6 generated the most realistic cost improvement curve to estimate future lot costs. We offered two fundamental rationales: (1) Lots 5 and 6 had incorporated much of the one-time cost reduction initiatives (CRIs), and (2) the configuration of units from Lots 5 and 6 was most consistent with that of Lots 7, 8, and 9. The cost of CRIs and configuration changes would be improperly extrapolated into the future if costs were based on data from Lots 1 through 6.

For the current analysis, we wanted to be as consistent as possible with earlier assumptions, but we found that we had to balance our previous arguments against new concerns that the methodology used to estimate post-gap production costs required us to estimate starting in Lot 1.

Using data from only Lots 5 and 6, or just Lot 6, was not desirable in the current analysis for two reasons. First, all the historical cases upon which the restart methodology is based use all data available. Our analysis of other programs that had been restarted used their entire program histories. Their original cost improvement curves were estimated using all historical information, not just recent lots. The change to the cost improvement curves after restart was based on these curves going back to Lot 1. We identified no reliable method to make historical cases comparable to the programmatic experience of the F-22A. Second, since (as we will see), a production gap shifts production far back along the cost improvement curve, getting production back to efficient levels will likely utilize the techniques developed by previous CRIs. Using data from just Lots 5 and 6 would underestimate likely cost improvement after restart.

To be consistent with our restart methodology and without any criterion to choose otherwise, we were left to utilize data for all of Lots 1 through 6 in forming estimates of future production cost.

Effects of Restarting the Production Line

We used the original F-22A MYP methodology estimate labor hours. To estimate hours after restart, the original C_1 (intercept) and b (slope) of the improvement curve were shifted by factors generated from analysis of a subset of aircraft restart data gathered by another RAND study (Birkler et al., 1993):[6]

$$C_{n\ \text{Post-Restart}} = X_1 \times n^{\ln(\beta)/\ln(2)} \times r^{\ln(c)/\ln(2)} \tag{3.2}$$

where

$\ln(\beta) = \ln(b) \times (-1.33949 \times \text{RUP} + 0.715537)$

$X_1 = \text{Max}(\text{MUN}, \text{RUP} \times \text{Q})^{\text{b}} \times \text{C}_1$ is the restart first unit cost

Q is the total quantity of aircraft procured before restart

RUP is retained unit percentage

MUN is minimum number of units retained.

RUP is the share of the total number of units produced before hiatus that appear to be retained at restart. RUP is determined by finding the unit on the original production cost curve that matches the cost at restart.[7] The formula determining β, the adjusted cost improvement slope, results from analysis of seven historical restart programs in which the estimated RUP was regressed on the ratios of estimated restart slope to estimated original slope. Table 3.2 includes outputs from Birkler et al., 1993. (See Appendix C for more details.)

6 The equation presented here is not found in the analysis by Birkler et al., 1993.

7 If 100 units of an aircraft were produced before a hiatus, and the production cost at restart is the same as the original unit 8, we say that the RUP is 8/100 = 8 percent.

Table 3.2
Retained Units Percentage

Type of Labor	Pessimistic	Average	Optimistic
Touch	0.8	13.6	38.1
Engineering	3.9	12.0	26.9

The formula for X_1, the estimated restart first unit labor hours (or material dollars), results from choices made in the model-selection process. It requires that the adjusted first unit hours be no lower than (a) the higher of an estimated minimal threshold (MUN or minimum unit number), or (b) the share of retained units times the number of units procured before restart. MUN is an estimate of the number of units that are considered to be automatically retained, implying that production will never restart with the equipment and processes originally used to procure the earliest units. For this study, a MUN of 18 for airframes was selected as appropriate, since this reflected the 17 engineering and manufacturing development (EMD) and production representative test vehicle (PRTV) airframes procured before lot production was under way. A MUN of 23 for engines reflects the 22 PRTV engines procured before lot production, but excludes EMD engine production.

To understand how Equations 3.1 and 3.2 are related and how the shift from one to the other can influence cost, Figure 3.1 shows how a notional unit cost improvement curve is modified by the restart methodology. The smooth line represents continued production, and begins to be dotted at the point of the shutdown decisions. The dotted portion can be thought of as representing the Continued Production and Warm scenarios. The bulleted line represents the Shutdown and Restart scenario: first unit cost after restart increases dramatically, and the slope is flatter; costs under the Shutdown and Restart scenario never become as low as if production had continued. (See Appendix B for more details.)

Figure 3.1
Notional Effect of Restart on Cost

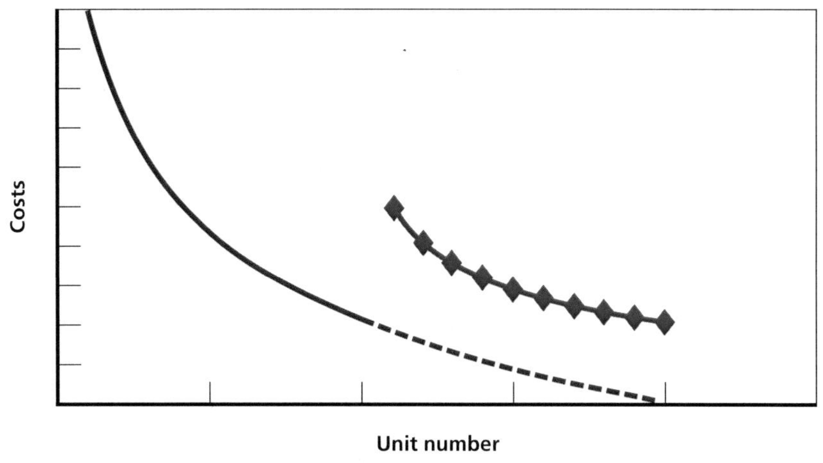

RAND *MG797-3.1*

Tail-Up and Last-Lot Rate Effects

In addition to standard procurement costs, we added a "tail-up" allowance of 7.4 percent to every last lot before hiatus or termination of production. This percentage was the agreed-upon upward cost shift for Lot 9 in the multiyear contract and was consistent with historical experience. However, we removed rate effects for the last lot of production, meaning that if 10 or 15 units were procured in the last lot of production before termination, the annual production rate effect was manually forced to look as though 20 units were being produced rather than the smaller quantity.[8]

Results of the Production Model

The results of the production model are presented in Table 3.3. Table 3.3 shows the sum of the annual TPC (airframe and avionics) and propulsion cost forecasts for Warm Production, Continued Production,

---

[8] For detailed analysis of tail-up costs, see Appendix B of Younossi et al., 2007.

Table 3.3
TPC and Propulsion Forecast (FY08 $million)

Scenario	2010	2011	2012	2013	2014	2015	2016	Total
Shutdown and Restart—pessimistic	—	—	1,695	1,930	2,803	2,489	2,517	11,434
Shutdown and Restart—average	—	—	1,578	1,922	2,885	2,613	2,674	11,671
Shutdown and Restart—optimistic	—	—	1,267	1,771	2,860	2,706	2,841	11,444
Warm Production	753	758	760	1,372	2,466	2,448	1,352	9,910
Continued Production	2,400	2,372	2,356	1,912	—	—	—	9,041

and all three variants of Shutdown and Restart.[9] It includes a 7.4 percent tail-up allowance: in 2013 for Continued Production, and in 2016 for Shutdown and Restart and Warm Production. The total TPC and propulsion for 75 aircraft in the Continued Production scenario is $9.0 billion; the Warm Production scenario, at $9.9 billion; the Shutdown and Restart scenario, from $11.4 billion to $11.7 billion. Focusing in on the Shutdown and Restart scenarios, we see that the considerable range from optimistic to pessimistic values seen in FY 2012 is entirely reversed by FY 2016, a result of how retained units interact with the cost improvement slope. It reinforces the point that these estimates should be considered approximations only.

Munitions, Support, Sustainment, Spares, and Logistics

While TPC and propulsion costs account for over 70 percent of the F-22A production budget, a substantial share of the budget is composed of what we will term "Other" goods (such as weapons, parts, whole engine spares, and trainers) and services (such as laboratories, engineering support, and logistics). As seen in Figure 3.2, the F-22A production budget since Lot 5 has been made up of roughly 59–65 percent TPC, 11–13 percent propulsion, and 22–29 percent other goods and services. In this section, we analyze in detail the history of the other goods and services procured by the F-22A production budget and forecast likely future costs.

The data presented in Figure 3.2 and in the rest of the analysis in this section are historical actual costs for Lots 5 through 7, combined with fixed-cost contractual information for Lots 8 and 9 from the F-22A program's 1537 budget justification.[10]

---

[9] We performed a sensitivity analysis of the RUP. The optimistic case, representing the highest RUP of the seven programs examined, and the pessimistic case, representing the lowest RUP, is shown in Birkler et al., 1993.

[10] "1537" is shorthand for Air Force Form 1537, required of all program offices by the programming process. This form structures the reporting of cost data into cost elements consistent with a program's work breakdown structure. When incorporated in the President's Budget, it is called the P-5 Exhibit. See Land, 2006, p. 14.

Figure 3.2
Percentage of the Annual F-22A Production Budget, by Major Item

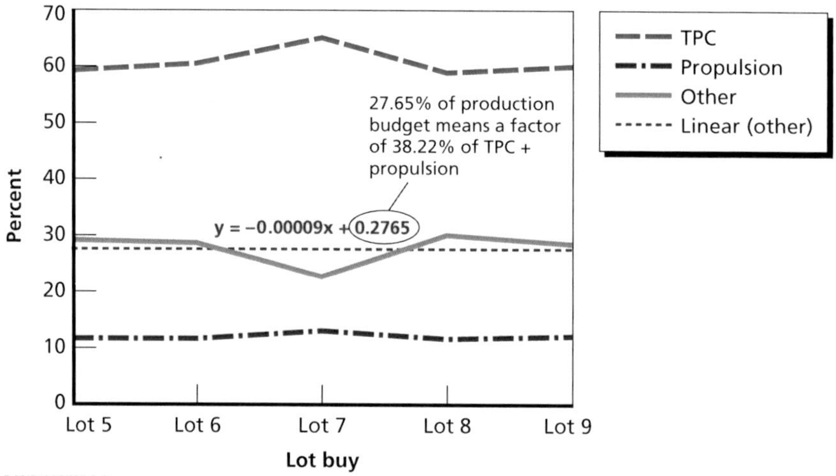

RAND *MG797-3.2*

We used a labor and material model to estimate TPC and propulsion costs, because there is a long tradition of using quantity-based models to make accurate predictions of unit and total airframe, avionics, and propulsion costs. Figure 3.2 suggests that if the program were stable, a simple, high-level estimate of Other costs could be made by multiplying TPC and propulsion costs by a factor derived from recent historical budget shares.

In fact, as seen in Figure 3.2, a line fit[11] through Lots 5–9 "Other" shares of the total budget is nearly flat, with an intercept of 27.65 percent, implying a total budget share of 27.65 percent. This, in turn, implies that a factor of 38 percent = [27.65 percent / (100 percent – 27.65 percent)][12] on TPC and propulsion would yield a reasonable, stable estimate of "Other" costs. While this is sensible methodology that produces a reasonable estimate for the Continued scenario, it does not provide guidance on how to modify the factor to estimate the costs

---

[11] Fit was made using ordinary least squares regression techniques.

[12] Since Other is defined by Other = (Total Budget – TPC – propulsion), Other/(TPC + Propulsion) = Other/(Total – Other).

Figure 5.2
Share of F-22 Vendors and Vendor Value, by System

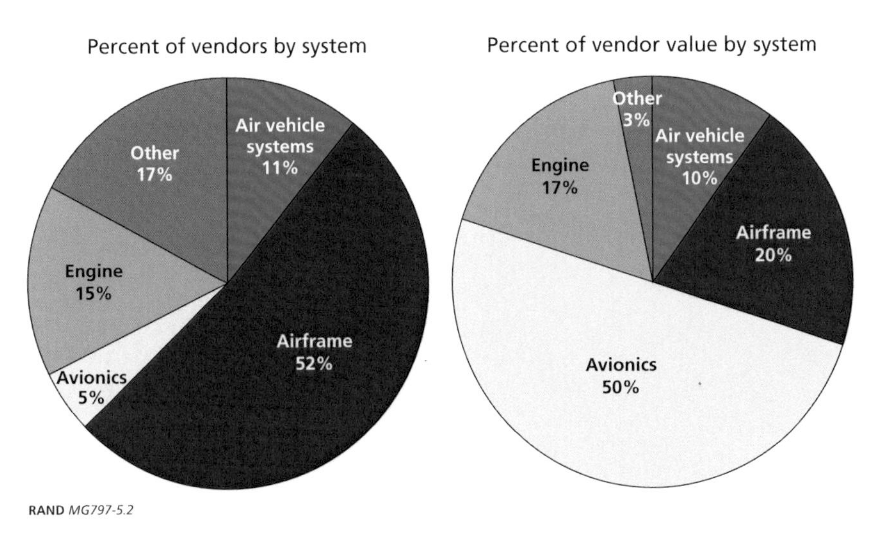

RAND *MG797-5.2*

In contrast to avionics suppliers, many more vendors work on the airframe, but the value of their contracts tends to be much smaller. Among all first-tier suppliers, 52 percent of vendors are estimated to contribute materials or services for production of the F-22A airframe while the engine and air vehicle systems account for 15 percent and 11 percent, respectively, of the remaining vendors. Vendors working on other aspects of the F-22A account for the remaining 17 percent of vendors but only 3 percent of vendor value.

F-22A Vendor Participation with the F-35 Program

Based on the information provided by the prime contractors, we estimate that approximately 63 percent of vendors involved with the F-22A program are also involved with the F-35 Joint Strike Fighter (JSF) program. Consequently, while JSF production is under way, a significant portion of the F- 22A industrial base is likely to continue performing related business activities regardless of the status of F-22A production. This will help sustain the F-22A industrial base in the event of a production gap.

That said, these vendors might also opt to commit resources (i.e., skilled labor, facilities, and tooling) that were previously dedicated to F-22A production to JSF production. Consequently, should production of the F-22A restart following a production hiatus, some vendors may find themselves unable to perform both JSF and F-22A activities simultaneously without significant additional investments. This could lead to cost increases and delays for one or both programs.

Assessment of Supplier Effects from a Production Gap

Managers at both Lockheed Martin and Boeing who interact with F-22A suppliers classified each vendor included in the vendor assessment sample as being at high, medium, or low risk of unavailability following an anticipated two-year production gap. Figure 5.3 provides a summary of the results extrapolated to the entire population of vendors involved with the F-22A program.

As a general guide, managers who completed the survey were told that high, medium, and low risk should correspond roughly to a 50 percent or greater, 10 to 50 percent, and less than 10 percent likelihood

Figure 5.3
Proportion of Vendors at Risk of at Least One Issue That Would Compromise Their Availability Following a Two-Year Production Gap

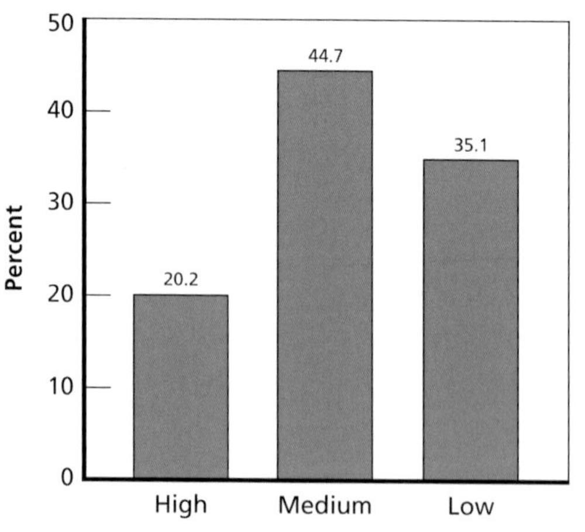

High: A 50 percent chance or greater of a vendor issue arising upon restart that would compromise their availability

Medium: Approximately 10 to 50 percent chance of a vendor issue arising upon restart that would compromise their availability

Low: Less than a 10 percent chance of an issue arising upon restart that would compromise their availability

RAND *MG797-5.3*

that an issue would compromise a vendor's availability following a two-year production gap, respectively.[1] As the respondents at the prime contractors noted, assessing the likelihood of the occurrence of an issue that might compromise vendor availability is difficult. As a result, these results should be viewed with caution.

Vendor Issues Caused by a Production Gap

To gain a better understanding of what is driving the prime contractors' concerns over vendor availability following a production gap, we asked them to characterize the likely causes of issues potentially leading to unavailability for each vendor in the sample. The issues cited include the possibility that vendors would

- discontinue business activities associated with F-22A production
- require a product redesign
- no longer have a process available that was previously required for F-22A production
- no longer employ people with the necessary skills
- have difficulty obtaining security clearances for their workforce
- no longer have facilities or tooling available that are needed for F-22A part production.

Figure 5.4 reports the results of this aspect of our analysis after extrapolating to the population of vendors working with Lockheed Martin and Boeing. Information provided by Pratt & Whitney would not support a comparable analysis, so its vendor base is not represented in this analysis. As the figure indicates, few vendors are viewed as being at risk of exiting the business; labor skill, process, and facility and tooling availability all ranked high in terms of concerns.

---

[1] To facilitate preparation of this study, a two-year gap was used in the quantitative analysis. If the gap turned out to be longer, the probabilities of vendor issues would increase, but we have no data to assess the size of the increase.

Figure 5.4
Concerns About Airframe Vendor Availability Following a Two-Year Production Gap for Lockheed Martin and Boeing Vendors

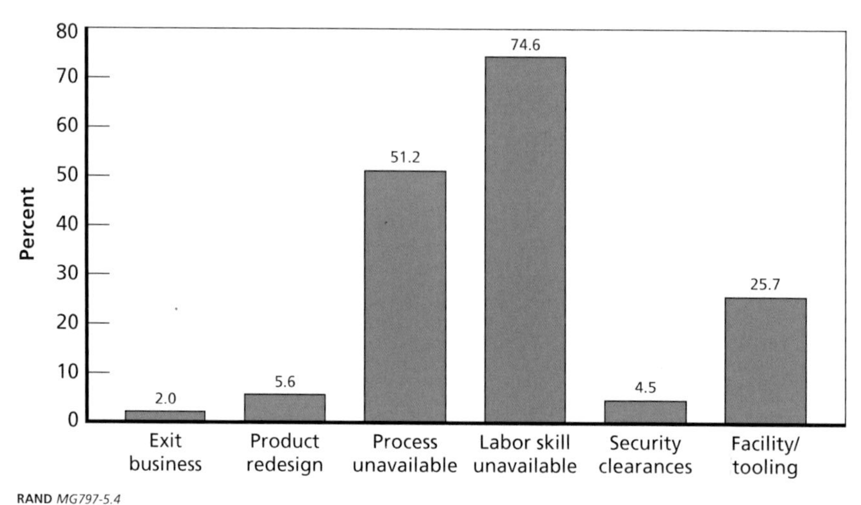

RAND *MG797-5.4*

Table 5.1 shows the percentage of vendors in different system categories that were classified as critical or at high risk of being unavailable following a two-year production gap. The table suggests that the greatest concerns pertain to avionic suppliers. These vendors tend to have the largest contract values due to the complexity associated with the systems they develop. The "Other" category ranks lowest.

Vendor Requalification

In the event of a production gap, it seems likely that some level of requalification would occur because of requirements necessitating a product redesign. Furthermore, vendor requalification issues were a specific area of concern noted by the prime contractors. As a result, RAND asked the prime contractors to indicate which vendors in their sample would be likely to face requalification issues following a two-year production gap. Lockheed Martin and Boeing's response when extrapolated to the population of vendors they work with suggests that nearly half of their suppliers are likely to face requalification issues in the event of a production gap. Pratt & Whitney indicated that vendor of production for the five-unit lot buys in the Warm Production and Shutdown and Restart scenarios, which reflect major shifts in production rate away from recent history.

The methodological solution we chose is to forecast individual line items from the 1537 program budget into a future full-production Lot 10. We forecast later full-production lots using inflation factors, and adjust line-item full-production lot estimates for changes in production rate. Only a limited number of line items apply when production is in hiatus or has been terminated. Before we present the estimate of the costs of these line items, however, we describe the work contained in each.

PALS

Performance-based Agile Logistics Support (PALS) consists of air vehicle spares, initial consumables of training equipment, and site activation spares.

Airframe Nonrecurring

Airframe nonrecurring cost includes DMS (described above), and refurbishment of tooling and special test equipment at vendors and prime contractors. In the past, it has also included unique identification tagging of inventory, as well as additional funding to ensure radar compatibility with air vehicle acceptance standards, but it will not include those elements after Lot 9.

Engine Nonrecurring

From Lots 1 through 5, substantial investments in Producibility Improvement Programs (PIPs) were made in engine producibility. No further PIPs are planned.

PSAS (PSS and PSP)

Program Support Annual Sustaining (PSAS) cost includes producibility and sustainability activities. PSAS measures the labor hours necessary to perform engineering support, customer support, tool manufacturing, manufacturing support, modeling and testing, quality assurance, and TPC plant engineering.

For analytic purposes, PSAS is broken into Program Support Sustaining (PSS) and Program Support Producibility (PSP). All PSP activities are production-related and depend on the quantity of aircraft being procured. We have interpreted program office research into the share of PSS that would be retained upon production termination to mean that roughly two-thirds of PSS activities are also production-related activities, leaving one-third to transfer to other budget activities upon termination.

PSAS has generally been estimated by F-22A contractor and program office personnel by a factor on TPC recurring hours, meaning that it is sensitive to production rate.

PSO

Program Support Other (PSO) consists of training equipment and devices, air vehicle peculiar support equipment, a field maintenance program, a structural retrofit program, aircraft mission equipment, a logistics tool program, and other miscellaneous support items. PSO tasks are estimated by the integrated program team based on material from Lockheed Martin subject matter experts. On a site-by-site, project-by-project, lot-by-lot basis, we estimated the number of people for the number of months required to complete specific tasks either by judgment or by historical experience with similar tasks.

OGC

OGC include additional GFE, equipment repair, chase support, laboratory infrastructure, common support equipment, outside studies, transportation costs, mission support, and other cost allocations and miscellaneous tasks.

Engine Support

Traditionally, engine support includes whole engine spares, engine support products, engine support services, field support and training (FS&T), and other modifications to engines incorporated into the production line. However, in our analysis, we considered whole engine spares as purchased along with engines in the production buy; costs for whole engine spares have been removed from the engine support category.

Useful Loads

These recurring flyaway items bought on a per-aircraft basis are purchased as part of a classified procurement program to support electronic warfare requirements.

In Table 3.4, we provide a history of all the Other costs found in the F-22A production budget and a forecast of what these other costs might be in Lot 10 under "full-rate" procurement of 20 aircraft. For all line items except PSS, the forecast is generated by a simple average of the Lots 5 through 9 expenditures. For PSS, Lot 6 included exceptional one-time costs, and its forecast was generated from Lots 7 through 9. Before putting too much confidence in these estimates, it is important to notice the large year-to-year variability in most individual line items, even under the relatively stable procurement quantities from Lots 3 through 9. This variability demonstrates a considerable change in work content from year to year within each line item; however, a large share of this variability is eliminated when line items are aggregated into a total Other cost.

In conventional military aircraft cost accounting, all the above items are usually thought of in two major types: (1) recurring and nonrecurring costs that, added to airframe, avionics, and propulsion, yield "flyaway" costs, and (2) nonrecurring and recurring below-the-line costs that, added to flyaway costs, yield total program costs.[13] We refer to these as *other flyaway elements*, and *below-the-line elements*, respectively.

---

[13] Total program costs include initial spares and additional BP-11 costs to yield total production costs; excluding them yields what is known as "weapon system" costs.

Table 3.4
History and Lot 10 Forecast of Other Flyaway and Below-the-Line Elements (FY08 $million)

	Lot									
	1	2	3	4	5	6	7	8	9	10
Aircraft quantity	10	13	21	22	24	23	20	20	20	20
Other flyaway elements										
Useful loads	0	0	6	29	27	43	35	33	33	34
Munitions	0	5	7	21	18	12	17	13	13	14
PSS (est.)	69	131	184	175	167	253	169	129	137	145
PSP (est.)	9	18	25	24	23	35	23	18	19	23
Airframe nonrecurring	226	559	542	352	172	138	98	67	94	114
Engine nonrecurring	25	53	66	18	4	0	0	0	0	1
Total	329	766	830	619	411	482	343	259	295	332
Below-the-line elements										
PSO	133	130	128	109	116	173	118	254	149	162
OGC	11	24	25	28	66	80	46	61	53	61
PALS	131	191	284	389	422	186	162	322	233	265
Engine support	88	125	91	120	167	83	64	155	186	131
Budget period 11	0	0	2	2	2	2	2	2	2	2
Total	363	469	530	648	774	525	392	794	623	622
Flyaway and below-the-line										
Total	692	1,235	1,360	1,267	1,185	1,006	734	1,053	918	954

To apply these line items consistently across all scenarios, we sorted the line items into three categories:[14]

1. Costs that accrue during production, during hiatus, and after termination (PALS, airframe nonrecurring, PSS)
2. Costs that accrue during production only but are insensitive to production rate (PSO, OGC, munitions, engine support)
3. Costs that accrue during production only but are sensitive to production rate (PSP, BP-11).

Segregating line items into these cost categories, while maintaining an understanding of which are flyaway costs and which are below the line, permits us to build total procurement estimates for all four scenarios while also maintaining the ability to generate flyaway cost figures.

Costs Incurred During Hiatus or After Termination

After production terminates or during a production hiatus, the lack of a production contract and budget means that certain goods, services, and support activities that are currently funded within the production budget, but are required to sustain the F-22A in operation, will transfer to F-22A sustainment and F-22A modernization programs.

To compare the total costs of the Shutdown, Shutdown and Restart, Warm Production, and Continued Production scenarios consistently from FY 2010 through FY 2016, we must explicitly include costs for these activities in all those years.[15]

First, we note that in our estimation, roughly two-thirds of PSS costs are production-related and will cease if the program goes on

---

[14] While each line item is itself an aggregate of innumerable goods and services, we have made certain that each of these is well defined enough to put into one of three categories.

[15] Federal agencies are appropriated and authorized funds by Congress. Once appropriated, funds can be obligated only on a prespecified category of goods and services over a set period of time: military construction for five years, procurement for three years, research and development for two years, and operations and maintenance for one year. Expenditures on those existing obligations can occur for up to five years after new obligations are no longer permitted. Details can be found in DoD Financial Management Regulation, 2008.

hiatus or is terminated. To obtain this fraction, we analyze PSS and PSP manning data for Lots 7, 8, and 9, as shown in Table 3.5.

The table shows two calculations. First, we found the share of PSAS dedicated to PSP; the remainder is PSS. Then, we found the share of PSS that is production-related (PSS/P) and the share that is sustainment-related (PSS/S). Overall, the ratio of PSP to PSAS manning is 12 percent. We assumed that the share of PSP costs is proportional to the share of PSP manning and applied the 12 percent share to generate the PSP estimate in Table 3.4.

We can now set PSP aside and focus on PSS. Unfortunately, only a portion of the PSS manning is provided by detailed element; the remainder is on a higher level. The detailed portion of PSS (921.8 of 1,350.6 full-time employees) is broken into six detailed cost elements: Production Development, Production Operations, Logistics Analysis, Program Operations, Quality, and Core Engineering (not shown). Production Development and Production Operations are grouped as PSS/P elements. Logistics Analysis, Program Operations, Quality, and Core Engineering are grouped as PSS/S elements. PSS/P elements will not continue during hiatus or after termination. However, PSS/S elements will continue under any scenario, even Shutdown. The share of PSS/S to all of PSS was estimated at 34 percent.

This 34 percent of PSS (roughly $49 million per year) will continue to be funded, regardless of the option chosen. Thus, roughly

Table 3.5
Suppport Workers Required for PSS and PSP

	Element	Total	Percent
Total	PSAS	1,532.8	
Total	PSS	1,350.6	88
	PSP	182.2	12
Detailed	PSS	921.8	
High-level		428.8	
Detailed	PSS/P	608.5	66
	PSS/S	313.3	34

$49 million × 7 = $343 million (FY08$) of the total cost of each scenario represents PSS/S activities that are currently funded in the production budget and that will transfer to the sustainment budget during hiatus or after termination. This money is included in all scenarios for all fiscal years, whether or not production continues. In Table 3.6, the $49 million can be clearly seen every year in the Shutdown scenario, and in FY 2014 through FY 2016 in the Continued Production scenario, although it is included in the cost of all other scenarios for all fiscal years.

The transfer of PSS/S to sustainment has considerable consequences for our cost estimates: for Shutdown, the cost of sustainment PSS over the FY 2010 to FY 2016 time frame ($342 million) is at least four times larger than the cost of terminating production ($79 million). This yields a total cost of $431 million for the Shutdown scenario. Also, this one-third of PSS is allocated in the Continued scenario from FY 2014 to FY 2016, totaling roughly $144 million.

In addition to the one-third of PSS costs, hiatus (Shutdown and Restart in FY 2010 and FY 2011) requires an additional approximately $114 million annually of airframe nonrecurring and engine nonrecurring costs. Although the airframe nonrecurring cost category is dominated by DMS (averaging approximately $75 million annually since Lot 5), it is also composed of other nonrecurring engineering activities, which we envision would be used as a DMS risk fund.

Extending the Forecast

Applying the Lot 10 forecast to all future lots within each scenario was a straightforward application of simple rules. The costs that depend on production quantity would be linearly interpolated from a full rate of 20, so that if 5 were being procured in a given year, 25 percent of the cost of that line item for that year would accrue. The costs that are fixed during production would show complete accrual regardless of production quantity. The costs that are accrued during a production hiatus would accrue, and so forth. The results for total flyaway additions and total below-the-line additions are presented in Tables 3.6 and 3.7.

Table 3.6
Additions to TPC and Propulsion Costs to Yield Flyaway Cost (FY08 $million)

Other Flyaway Elements	2010	2011	2012	2013	2014	2015	2016	Total
Shutdown	49	49	49	49	49	49	49	345
Shutdown and Restart—pessimistic	164	164	182	223	332	332	332	1,729
Shutdown and Restart—average	164	164	182	223	332	332	332	1,729
Shutdown and Restart—optimistic	164	164	182	223	332	332	332	1,729
Warm Production	182	182	182	223	332	332	223	1,657
Continued Production	332	332	332	278	49	49	49	1,421

NOTE: Totals do not sum due to rounding.

Table 3.7
Additions to Flyaway Cost to Yield Total Cost (FY08 $million)

Below-the-Line Elements	2010	2011	2012	2013	2014	2015	2016	Total
Shutdown	—	—	—	—	—	—	—	—
Shutdown and Restart—pessimistic	—	—	622	622	622	622	622	3,108
Shutdown and Restart—average	—	—	622	622	622	622	622	3,108
Shutdown and Restart—optimistic	—	—	622	622	622	622	622	3,108
Warm Production	622	622	622	622	622	622	622	4,351
Continued Production	622	622	622	622	—	—	—	2,486

NOTE: Totals do not sum due to rounding.

Summary

In this section, we aggregate the costs of termination, hiatus, restart, and production, and compare them across scenarios. Table 3.8 presents five of the most commonly used cost comparisons. In the second column are the sums of all hiatus, restart, and termination costs for each scenario. In the third column are flyaway unit costs, which are the sum of TPC, propulsion, and other flyaway cost elements, divided by 75; these range from a low of $139 million for Continued Production, to $154 million for Warm Production to $179 million for Shutdown and Restart. The fourth column is average unit cost (AUC), which includes all costs—hiatus, restart, termination, procurement, other flyaway, and below-the-line—divided by 75.[16] In the fifth column is total cost, which is AUC multiplied by 75, except for the Shutdown scenario, in which the $0.4 billion cannot be spread over any aircraft.[17] The final column also contains the total cost, but in then-year dollars. While the

Table 3.8
Total Costs, by Scenario (FY08$)

Options (2010–2016)	Hiatus, Restart, and Termination Costs ($million)	Flyaway Unit Costs ($million)	Average Unit Cost ($million)	Total Cost ($billion)	Total Cost (TY $billion)
Shutdown	79	—	—	0.4	0.5
Shutdown and Restart	513	179	227	17.0	19.2
Warm Production	111	154	213	16.0	17.8
Continued Production	79	139	173	13.0	13.8

[16] Note that AUC is for the next 75 units only. This should be clearly distinguished from average procurement unit cost (APUC) and program acquisition unit cost (PAUC), both of which are calculations of cost of all units procured since the start of a production run, which would in this case include the 183 F-22A already procured plus the next 75.

[17] The $0.4 billion for the Shutdown scenario includes $79 million estimated for termination. The rest of the $0.4 billion is for sustainment costs for aircraft currently under production that are transferred to other F-22 programs upon termination.

order of options from least to most costly does not change, putting our estimates in then-year dollars expands the range of costs considerably, because the recurring costs for Shutdown and Restart are further out into the future than for either Warm or Continued Production.

CHAPTER FOUR

Sustainment, Modernization, Technical Data Package, and Contract Closeout

This chapter addresses four issues that could be affected by a change to the currently planned program of record for the F-22A. These issues are

- F-22A sustainment efforts (logistics support)
- the F-22A modernization program
- technical data package delivery to the USAF
- contract closeout of the F-22A production contracts.

We analyze each of the four production options described previously in connection with each of these four issues.[1]

Sustainment

In 1995, the USAF decided that contractors should provide off-equipment[2] sustainment support for the F-22A: the Lockheed Martin/Boeing team for the air vehicle, training systems, and support equip-

[1] To reiterate the program of record used in this analysis: Production ends with 183 aircraft after Lot 9 is complete; the last delivery occurs in 2011; shutdown activities begin in FY 2009 for some parts of the production line; the sustainment program is based on 183 aircraft; and the modernization program is based on 183 aircraft.

[2] *Off-equipment* refers to the majority of sustainment activities that are not performed on the aircraft itself on or near the flight line. Thus, supply chain management, inventory control, management of spare parts repairs, sustaining engineering, etc., will be the responsibil-

ment; and Pratt & Whitney for F119 engine-related sustainment. That concept is part of the current program baseline through 2012, although it is possible that the USAF will revisit the costs and benefits of retaining all currently planned sustainment roles with the contractors through a business case analysis study which is currently in progress. However, the current F-22A sustainment program consists of two major contracts awarded in early 2008: the Follow-on Agile Support To the Raptor (FASTeR) contract awarded to the Lockheed Martin/ Boeing team, and the Sustainment Program for the Raptor Engine (SPaRE) contract awarded to Pratt & Whitney. Both contracts continue support provided under earlier sustainment contracts.

The baseline for both contracts will evolve over the next four years as aircraft are delivered as part of Lots 7–9 under the multiyear contract. By 2012, assuming further additional procurement after the current multiyear production for Lots 7–9 has not been approved, these sustainment activities will be in a relatively steady state, with all aircraft delivered to the operational commands. Contractor and USAF personnel requirements should be stabilized as experience is gained through the accumulation of flying hours, spare parts repair requirements become better known, and consumable supply use can be more accurately determined.[3]

For Option 1 (Shutdown with no intention of restarting), no change will occur to the planned sustainment activities for the 183 aircraft. As production is shut down, some engineering and other support will transition from the production contracts to the sustainment contracts.[4] With no more production, the sustainment and modernization contracts will have to fund all contractor engineering and other

ity of the contractors. On-equipment maintenance will be performed primarily by USAF military personnel.

[3] A major milestone in the maturity of an aircraft is the accumulation of 100,000 flying hours. At that point, the experience of the USAF in operating the aircraft enables better predictions of maintenance workload, break rates, reparable spare part repair data, consumption of supplies, etc. As currently forecast, the F-22 should reach that milestone in 2010.

[4] For F-22 manufacturing, two production contracts are developed: one for the Lockheed Martin team for the air vehicle and associated support and one for Pratt & Whitney for the F119 engines and associated support.

support required. Replenishment spares will be ordered only for sustainment requirements since there will be no further production orders for parts, resulting in smaller buys of individual parts and consumable supplies, probably at higher costs due to the smaller orders. Alternate sources of supply will have to be found for some parts in conjunction with DMS activities.[5] Depot maintenance activities will continue at USAF air logistics centers (ALCs), Lockheed Martin's facility at Palmdale, California, and some original equipment manufacturers (OEMs) (for spares repair not performed at USAF ALCs). Production data, aside from that needed for supporting the reprocurement of spares, would be stored but not regularly updated (if at all). The USAF and Lockheed Martin may agree to retain some of the production and test tooling for later use by sustainment activities, but all repair activities and repair data will basically be in place at the ALCs and at Palmdale by the end of the currently planned production. Although all the planned sustainment activities may not be fully funded in the USAF Program Objective Memorandum (POM) and budget, no additional sustainment requirements that are not part of the program of record should be required under Option 1.

Under the Shutdown and Restart option (Option 2), most of the activities discussed above will still occur. During the interim shutdown period of two years, the USAF and contractors may decide to add personnel previously working on production (categorized under the PSS portion of the production contracts) to the sustainment contracts to keep essential skills in place during the interim transition period. This would depend upon whether some sort of production "bridge" contract could be crafted to address these interim activities, or whether the sustainment and modernization contracts would be the vehicles for these interim activities. Production data would have to be kept updated as modifications were approved for the F-22A, so production could restart with up-to-date data packages.

Assuming a bridge contract is not established for the interim shutdown period, the sustainment contract work scope and funding would undoubtedly increase to cover personnel whose skills would be required

---

5 See discussion below on DMS.

for restarting production. The exact number is difficult to estimate and would involve much negotiation between the USAF and contractors as to exactly who should be retained. However, after production restarted, the sustainment contract should return to its baseline level for a couple of years until additional aircraft deliveries begin. It may be possible that the sustainment contractor requirements could dip below the baseline level somewhat as production personnel were "shared" again with the sustainment efforts. In addition, prices paid for purchases of replenishment spares and consumables could be lower as production and sustainment purchase quantities were combined, resulting in lower unit costs. As the additional aircraft are delivered to the operational USAF, additional sustainment costs (such as for management of spares, depot maintenance, and consumables, additional support equipment, etc.) would be incurred for a larger fleet of aircraft.[6] In addition, key decisions would be required as to the size of the additional F-22A buy and where the aircraft would be based.[7] If any additional bases were to be converted to F-22As, the sustainment contract would have to reflect one-time and recurring contractor costs for supporting the additional F-22A locations. After the second production, the baseline would be much as described above for the production shutdown option, but at a higher level of effort depending on the size of the buy and the basing option chosen.

Under the Warm Production option (Option 3), some cost avoidance to the baseline sustainment program could result from sharing production personnel and combining buys of production and replenishment parts and consumables. Increases to sustainment, however, would result from increased requirements for a larger operational fleet

---

[6] In addition to sustainment contract costs, the USAF would also incur additional costs for additional USAF operational and sustainment personnel, as well as fuel, depot level reparables, and consumable supplies, which would be purchased from the Defense Logistics Agency.

[7] Depending on the size of the follow-on buy, the USAF would have several options for beddown, including increasing the size of the current F-22A squadrons, adding squadrons to existing F-22A bases, or creating new operating locations for the F-22A, either at active, Air National Guard, or Air Force Reserve bases.

(spares, consumables, fuel, depot maintenance, USAF personnel, support equipment, etc.) and potentially higher basing costs.

Finally, under Continued Production (Option 4), some cost avoidance for sustainment could be forecast due to sharing of production personnel and merged buys of production and replenishment spares. Increases, of course, would be expected for additional aircraft being added to the operational fleet as described above.

Modernization

Because the F-22A is a modern, state-of-the-art weapon system, it has an active program to ensure that its capabilities are kept current with known threats. As threats evolve, a team works out a solution using newer technologies, the affected equipment is modified, and the solution (either a hardware or software modification) is incorporated as quickly as possible into production aircraft. Because the F-22A has been in production throughout the decade, many of the previously produced aircraft will have to be "retrofitted" at some point with these newly developed capabilities. Thus, in addition to the engineering design efforts, there are direct labor activities involved with removing old equipment and installing the latest hardware, normally when each aircraft returns to one of the two depots (Palmdale, California, or Ogden, Utah). Software modifications are handled as a continuous process of updating all aircraft by uploading the latest software as quickly as possible after release of a particular update. This can generally be performed wherever the aircraft are located.

A second aspect of the modernization program, worked on hand-in-hand with sustainment efforts, concerns DMS activities. Because of the evolution of commercial technology (especially information technology) and the relatively small market for F-22A parts (almost all of which are unique to the F-22A), some manufacturers may discontinue manufacturing of these parts from time to time. This leaves the USAF with a decision—buy as many parts as possible for future use in either production or sustainment, find another manufacturer who will make an identical part, or redesign the part using the latest technology that the industry's manufacturing capabilities will support for at least the foreseeable future. In some cases, the Air Force might be able to design and manufacture certain parts in a depot. Having a manufacturer produce the number of parts forecast to be needed for several years may be a short-term solution, but funding availability, procurement restrictions, or uncertainty about other system design changes may constrain this option for many parts. Because of the integrated nature of the various F-22A systems and subsystems, introducing new parts often requires reengineering a complete system or subsystem to make it compatible with the new technology, unless the new part is of the "plug-and-play" variety. Thus, there is a continuous effort by the USAF and the contractors to ensure that parts are available for production and sustainment.

Because of DMS issues, the F-22A modernization program has a baseline of activities, much like the sustainment activities discussed above. As currently designed, the modernization program will incorporate modifications into aircraft as they are produced through Lot 9 of the MYP contract. Those not updated during MYP production will be updated afterward, as will earlier aircraft.

The modernization baseline program will not change if production is discontinued after Lot 9. Threat analyses, engineering efforts, and aircraft modifications will continue as new enemy capabilities emerge and available funding permits. As production ends, personnel will transition from production to modernization activities exclusively, because the possibility of sharing personnel between the production and modernization programs disappears after shutdown.

Under the Shutdown and Restart option, modernization retrofits and threat update work will continue. Unless a "bridge" production contract can be developed and funded, some production personnel involved with threat updates or DMS work will be funded as part of either the FASTeR or FS&T sustainment contracts or the modernization contracts if their expertise is to be retained, thereby increasing requirements and required funding for modernization (at least during the two-year gap). After production is restarted, some of these people could return to the production contract or could be shared between the two. Much like sustainment, modernization equipment costs will increase due to the larger fleet, but some retrofit costs may be reduced if new equipment can be incorporated into aircraft on the production line, thereby avoiding installing the older equipment during production and then having to replace it later with a newer version. In addition, modification kits for earlier aircraft can be combined with production buys to help reduce unit costs. Because of the production gap, some DMS manufacturers who otherwise might continue producing obsolete technologies under continuous production may decide to discontinue production of F-22A parts or to drop out of defense work entirely. Quantifying the amount of additional DMS workload would be extremely difficult to forecast because part of each supplier's decision to resume manufacturing the same F-22A part or equipment would hinge on its perception of the likelihood of a restart and its commitment to continue producing what may be commercially obsolete technology.

Both the Warm Production option and the Continued Production options have the same effect on modernization. Engineering talent could continue to be shared between production and modernization, thereby producing a slight reduction to baseline modernization requirements while production is under way. Both might allow the combining of modification kits (parts or systems) with production buys for new aircraft, thereby reducing unit costs. Both Warm Production and Continued Production could produce somewhat higher downstream costs due to a larger fleet size than the 183 aircraft baseline. In addition, continued production could result in higher retrofit costs later in the program (even with the same number of total aircraft as under Warm Production) because future updates would have to be incorporated after a larger number of aircraft had been produced.

Technical Data Package

Aircraft technical data is a widely used term with many different meanings. In the context of this study, technical data consist of three data categories: (1) design and production, (2) repair, and (3) usage.

The first category includes automated design tool data (for example, computer-aided three-dimensional interactive application [CATIA] data), manufacturing data derived from that design data, lists of parts required to build an aircraft, data used to program automated production tools, and assembly process data. As an aircraft production line is being shut down, a key question that must be addressed is how much design and manufacturing data should be retained.

After that question is answered, a follow-on question is whether the data should be retained as of the last production lot or whether activity should be funded to update design and production data with changes made later as part of the sustainment or modernization programs. If there is any likelihood that a production line will be reopened for a part or system, the storage of the data for future use must be carefully analyzed. Despite the best efforts to document design or production data at the end of production, knowledgeable people who have some corporate memory are needed to make the data usable, unless an active update program (thereby keeping people familiar with the data) has been continued after production. Another consideration for the government in retaining design or manufacturing data is whether they will ever want to recompete production of a part or even a system in the event that an OEM's price is viewed as being excessive. Without design and manufacturing data, an alternate source of supply would be required to reverse-engineer a part in order to manufacture it.

A second type of data is repair data. These consist of detailed instructions required to repair a reparable part, normally at a government depot, commercial company specializing in part repair, or the OEM's facility. Repair data would normally not be as detailed as design data, so manufacturing a part would normally not be possible with only repair data. Under the current sustainment program, the USAF ALCs will repair most F-22A reparable parts, with some going back to OEMs for repair. This is being done under a partnership with the Lockheed Martin/Boeing team for the air vehicle and Pratt & Whitney for the F119 engine. Those repair capabilities are scheduled to be in place by 2010, except for software maintenance. Thus, whatever technical repair data are required will be in place before the last MYP aircraft is delivered as part of the sustainment baseline program, except for software.

So no additional costs for repair data should be incurred, regardless of any production decision made.

The third type of technical data is usage data, e.g., aircraft break rates, times between repair, time to repair, and consumption of supplies. Currently, these data are gathered and analyzed by the contractors as part of the sustainment contracts, and this is expected to continue, so these data will be available to the USAF as needed.

Thus, the question arises as to what costs the USAF will incur for delivery of a technical data package. As noted above, repair instructions and usage data are already part of the baseline sustainment program, so only design and production data need to be considered. The USAF will have to decide what needs to be retained for future use and what can be stored (and perhaps forgotten).

In terms of the four production options, the question is not cost, but rather timing. While production continues, design and production data will be maintained and updated by the contractors. When production finally ends, the USAF will have to decide the disposition of the design and production data (retain all, some, or none). The overall effort will essentially be the same, whether the decision is made after Lot 9 or whether production continues for several more lots. So, in terms of our analysis, aside from the effects of inflation during the years of extended production, the basic effort for producing whatever the USAF needs as a technical data package will be the same. Thus, it becomes a matter of funding these efforts in the proper year(s) and determining what the data package will entail.

Contract Closeout

After all efforts or items required by a contract have been completed or delivered, a series of administrative actions is necessary to satisfy the government that all obligations have been met and disputes have been settled. At the end of these actions, a final payment is made, and the contract is closed out. Contract closeout involves a large number of people from both the government and contractor, especially when a contract involves several years of effort (several delivery lots such as under the MYP) and a large amount of activity, such as the complex manufacturing process for the F-22A.[8] Government personnel involved would include the administrative contracting officer, people from the Defense Contract Management Agency, Defense Finance and Accounting Service, Defense Contract Audit Agency, the contractor(s), and perhaps other government agencies, depending upon the issues encountered. FAR part 4.804-1 provides guidelines for the time standards (maximum time) within which contract closeout should be completed after the contracting officer is notified that all deliveries have been completed. The maximum time allowed is 36 months for contracts involving indirect cost rates; other contracts, such as fixed price contracts, are expected to be closed within six months of final delivery notification to the contracting officer.

In the case of the F-22A, each production lot has been placed on a separate contract for Lots 1 through 6. For Lots 7 through 9, one MYP contract will be used for all 60 aircraft and another for the associated engines in the multiyear buy. Thus, in the case of production shutdown, contract closeout procedures will be initiated after delivery of the last aircraft and all other requirements have been met. Whether the contractor charges these efforts as an overhead expense or directly to the F-22 MYP contract would not change their cost, regardless of what decision might be made concerning additional aircraft post-MYP.

In the case of Shutdown and Restart, a new production contract would be initiated when production is resumed. After deliveries are completed, contract closeout procedures would be performed as described above.

The same procedures would be accomplished for however many additional lots would be procured beyond the multiyear buy. In the event of a decision to procure additional aircraft under a new multiyear contract, lots could be bundled into one contract. Thus, in terms of cost, there would be no change to the baseline costs for Lots 7–9 (whether a decision was made to fund them as an overhead cost or a direct charge), and each new contract initiated would incur its own

[8] Federal Acquisition Regulation (FAR) Part 4.804-5 provides detailed procedures required for closing out contracts.

closeout costs, whether under Warm Production or Continued Production. There would be a difference if the number of contracts was not the same for each option.

Summary

In terms of costs to the F-22A program, Shutdown would produce no additional costs above the baseline program of record in any of the four areas. This does not imply that all these efforts have been fully funded in the USAF POM or budget; rather it indicates that requirements should not change due to a decision not to continue production after Lot 9, the last lot of the multiyear buy.

Under Shutdown and Restart, sustainment costs would increase somewhat during the production gap due to the migration of "banked" production personnel (assuming the absence of a bridge contract) and would then experience some cost avoidance as personnel and spares buys are shared or merged with production requirements when production resumes. As aircraft deliveries are made, overall sustainment costs would increase due to additional flying hours and fleet size and perhaps additional basing costs. Effects on the modernization program parallel the sustainment program: increases due to banked personnel during the gap, some reduction after the production restart due to sharing of personnel and merged buys of parts and kits with production, and increases in retrofits as fleet size increased. One unique effect on the modernization program would be the effect of the gap on DMS, with the gap potentially exacerbating an ongoing effort to solve DMS issues. The technical data package costs would be unaffected by this option compared to the baseline, except for the effects of inflation. Contract closeout costs would increase as part of each additional production awarded.

Warm Production and Continued Production have nearly identical influences, with some shift in timing for certain items. Sustainment would experience some cost avoidance during the active production years as personnel were shared and spares buys merged with production. At the end of production, sustainment would increase as personnel transitioned to sustainment activities and spare parts prices increased due to smaller buys for just sustainment. In addition, due to larger aircraft inventories, overall sustainment costs would rise. The choice of how to base additional aircraft could create further nonrecurring and recurring costs. Modernization would be much like sustainment, experiencing some cost avoidance during active production and then somewhat higher costs for post-production personnel migration and reduced parts buys. DMS would basically be the same under either of these two options. Retrofit activities would increase with the larger inventory of aircraft. The cost of the technical data package would essentially be the same; inflation would be the only difference, depending on when the USAF wanted data to be delivered. Contract closeout would again be a unique cost of each production contract awarded.

CHAPTER FIVE

The Effect of the Production Gap on Vendors

Over time, defense prime contractors have moved toward a role of system integrator, relying heavily on other firms to fabricate subsystems and components. The F-22A is a leading example of this trend, with hundreds of vendors contributing products and services to a complex production process coordinated by three prime contractors: Lockheed Martin, Boeing, and Pratt & Whitney.

In this chapter, we summarize findings from a vendor assessment survey that was issued to the prime contractors to gather information on the industrial base to understand their perceptions of how a production gap would affect the vendors involved with the F-22A program. The F-22A airframe prime contractors were reluctant to allow RAND to directly survey their vendors because they ascertained that the survey would make future negotiations between the prime contractor and the vendors difficult.

The Vendor Assessment Survey

The vendor assessment survey requested information on the business activities of firms supporting each of the prime contractors. In addition, managers who interact with suppliers at each of the prime contractors were asked to provide a qualitative assessment of the risks and problems that might be experienced by vendors supporting F-22A production in the event of a production gap.

The survey requested information on the first-tier vendors because the prime contractors have limited knowledge of the business activities of lower-tier vendors. Consequently, estimates discussed throughout the rest of this chapter are representative of only the first tier; we did not collect data that would enable us to draw inferences about how a gap might affect lower-tier vendors.

Sampling Approach

To make completion of the vendor assessment survey manageable, RAND requested information for a sample of vendors working with Lockheed Martin and Boeing. The samples for each of the non-engine prime contractors were developed independently and according to a stratification approach. Under this approach, the population of first-tier vendors contracting with each of the prime contractors was divided into two groups (or strata) based on each vendor's contract value from the Lots 7–9 multiyear contract. A census of the 50 largest vendors in terms of contract value was included in the first stratum of each prime contractor's vendor sample. The second stratum was made up of a random sample of 50 of the remaining vendors with smaller contract values. The advantage of such a stratification approach is that we could obtain more information on critical vendors. For instance, even though our sample includes only 17 percent of Lockheed Martin's vendor base, these vendors represent over 90 percent of the total vendor value associated with Lockheed Martin's share of the multiyear contract.

Pratt & Whitney contracts with fewer vendors than do Lockheed Martin and Boeing. Consequently, it was feasible to sample all its first-tier vendors. As a result, we have greater confidence in our vendor estimates for Pratt & Whitney.

Because we used a stratified sampling approach for two of the three prime contractors, care must be taken when extrapolating results to the population of first-tier vendors. The standard approach for extrapolating results from a stratified sample involves assigning each vendor in the sample a sample weight based on its probability of being included in the sample. For the analysis, we constructed sample weights and used them to obtain population estimates.

Caveats

While our approach is structured in such a way that we believe it provides important insight about the F-22A industrial base, we should make some caveats. First, because nearly all the information we rely upon is proprietary, verifying its accuracy is difficult. Second, some of the findings presented here are based on subjective assessments made by managers at the prime contractors, and this can make comparing responses between prime contractors difficult. Furthermore, there is a concern over whether the assessments are truly objective. To address this concern, we asked the survey respondents to elaborate and justify some aspects of their assessment. Third, in some instances, the prime contractors provided different levels of detail in their responses, and for various reasons data were not always available for every vendor included in the sample. Finally, we focused on vendors that currently support F-22A production. Should some vendors face issues that will affect their ability to support F-22A production in the future, other suppliers not currently supporting production may be engaged. As a result, some of the effects on the F-22A program that are identified here may be mitigated by contracting with new suppliers.

All these issues can create difficulties when comparing the survey responses provided by each of the primes. We have dealt with these issues to the best of our ability and noted areas where concerns over these types of issues remain.

The F-22A Industrial Base

Lockheed Martin is the leading firm in charge of all aspects of F-22A production except manufacturing the F119 engine. It performs activities associated with about 25 percent of the non-engine contract value through its various business divisions, while Boeing's internal share is estimated at around 12 percent (King and Driessnack, 2007) (Figure 5.1). The remaining 63 percent of manufacturing of the airframe and integrated systems is carried out by suppliers. Pratt & Whitney manages manufacturing of the F119 engine and contracts with various vendors. These vendors make up a similar share of the engine contract value.

Figure 5.1
Structure of the F-22A Industrial Base (Excluding F119 Engine Production)

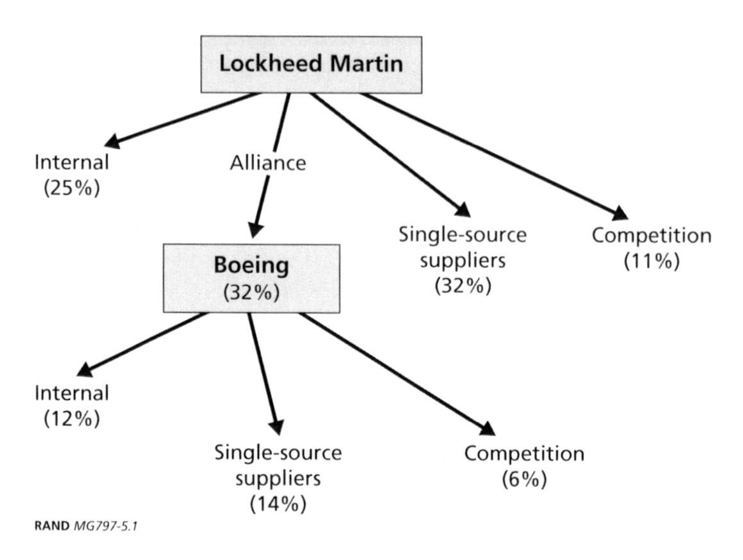

In general, the prime contractors use competitive procurement for components and materials that involve limited complexity and are available from multiple sources (King and Driessnack, 2007). While the prime contractors are in some cases able to call on multiple suppliers for parts and materials, they have developed strategic alliances and long-term contracts with more than one supplier of parts sourced on a competitive basis, in addition to long-term agreements with single-source suppliers.

Of the nearly 1,000 suppliers working on the F-22A, we estimate that approximately 5 percent (50 vendors) work on avionic systems (Figure 5.2). While these firms represent a small segment of the vendor base, because of the complexity and importance of the systems they provide, they account for a much larger share of the total cost of the F-22A. British Aerospace (BAE), Northrop Grumman, and Raytheon hold the largest vendor contracts with Lockheed Martin, and all supply avionic systems. Among Boeing suppliers, Northrop Grumman and Raytheon have teamed to provide the radar system, which is the largest vendor contract among Boeing suppliers.

Collate Color
Text
Section

Collate Color Text Section

Table 5.1
Critical and At-Risk Vendors, by System

	Number of Vendors	Percent Critical to the F-22 Program	Percent at High Risk of Having an Issue Compromising Availability After Production Gap
Air vehicle systems	106	45.4	39.5
Airframe	525	34.0	16.5
Avionics	54	79.5	44.1
Engine	146	NA[a]	24.3
Other	167	8.1	8.1
Total	998	33.2[b]	20.2

[a] The prime contractors did not provide data for critical engine vendors.
[b] Excludes engine vendors.

requalification issues would be less severe, with fewer than 5 percent of its vendors characterized as being at risk of requalification issues. However, should the production gap last longer than two years, Pratt & Whitney anticipates that a much larger share of vendors will require requalification.

Requalification is an important issue because it can create delay and inflate restart costs. According to the prime contractors, it typically takes up to six months for a current vendor to go through requalification of an existing process. If a current supplier needs to qualify a new process, the likely time required to qualify grows to between six months and a year. If a new source must be engaged to develop a complex system, the qualification process can take up to five years and cost many millions of dollars.

Conclusion

The F-22A program has developed unique fighter technologies. Inevitably, other military aircraft, such as the F-35, have benefited from these advancements. In cases where the F-22A uniquely employs advanced technologies or manufacturing processes, either temporary or permanent discontinuance of production (Shutdown and Restart and Shutdown, respectively) will likely cause the loss of some of the skills and knowledge used to produce the F-22A as firms refocus their business activities on other areas.

Our analysis suggests that while few vendors are expected to go out of business as a result of a two-year production gap, other issues are likely to hinder the program's restart capabilities, should further production be authorized. These include issues associated with requalifying the vendor base as well as concerns over the availability of skilled labor, processes, facilities, and tooling used by firms supporting F-22A production. All these issues are likely to increase the occurrence of DMS issues.

CHAPTER SIX

Summary and Policy Options for the Air Force

This chapter summarizes our findings regarding options for the F-22A industrial capability.

Cost Estimates[1]

This section outlines the cost implication of each of the alternatives. All these options will cost money, and Table 6.1 summarizes our cost estimates for them. The first column includes the shutdown costs for all the options as well as the costs of a production hiatus, restart, and termination for the Shutdown and Restart option. The second, third, fourth, and fifth columns compare the hiatus, flyaway unit costs, average unit costs, total cost in FY 2008 constant dollars, and total cost in then-year dollars, respectively.

---

[1] An April 2009 Air Force review of the draft monograph suggested that the estimate of the tooling required to be stored had risen, and that these higher estimates would have increased the cost estimates. We were unable to validate the new data and thus retained the estimates we had previously obtained from contractor and government sources and validated.

Table 6.1
Total Cost, by Scenario (FY08$)

Options (2010–2016)	Hiatus, Restart, and Termination Cost ($million)	Flyaway Unit Cost ($million)	Average Unit Cost ($million)	Total Cost ($billion)	Total Cost (TY $billion)
Shutdown	79	—	—	0.4	0.5
Shutdown and Restart	513	179	227	17.0	19.2
Warm Production	111	154	213	16.0	17.8
Continued Production	79	139	173	13.0	13.8

NOTE: The average flyway unit cost and the AUC are based on 75 aircraft.

Options for the Air Force

We analyzed four options: Shutdown, Shutdown and Restart, Warm Production, and Continued Production at the current rate of 20 aircraft per year.

Shutdown

Upon shutdown of the F-22A production line, all the suppliers, including the prime contractors, will shut down their production lines permanently, and tooling and equipment needed for the production of F-22A airframe, engines, and other related components will be disposed of. This option likely will cost about $79 million. However, if the Air Force plans on future modernization, upgrade, or a service life extension program, it should decide the scope of those activities and determine what tools, equipment, and technical data must be saved. This option would undoubtedly require additional funding not currently included in our shutdown cost estimate.

Shutdown and Restart

The option of temporarily halting F-22A production entails closing the production line for a period of time and storing all related tooling, equipment, facilities, and technical information in such a way that the production line could be reopened with ease. For a two-year shutdown followed by production of 75 aircraft, the likely cost estimate of this option is about $513 million. This includes $434 million for the initial shutdown and restart and an additional $79 million for the final shutdown once the restart program concludes the delivery of the last aircraft. This option requires congressional authorization for additional funds so the Air Force can manage at-risk suppliers that would otherwise need to be requalified. Even though this option preserves the production capability, producing aircraft after a production hiatus would cost, on average, 40 percent more than continued production of 75 units.

Warm Production

Under this option, the production line remains open by procuring three lots of five aircraft each, for a total of 15; then ramping up to procure the additional 60 aircraft. The 15 aircraft would cost more than double what the Air Force is paying for the F-22A under the current multiyear contract, because of the low production rate. To mitigate a production shutdown of many suppliers, advance procurement funding is required in the near term. This option keeps the production line open, but producing those 75 aircraft would cost, on average, 30 percent more than continued production at the current rate of 20 aircraft per year.

Continued Production

If the Air Force were to continue to procure F-22As at the same rate as the current multiyear contract (20 aircraft per year) then there are no additional nonrecurring costs. As with Warm Production, advance procurement funding would be needed to mitigate the possibility of a production shutdown at many suppliers. The average unit cost of the next 75 units would remain about the same as for those bought during the MYP.

Sustainment and Modernization Program

Shutdown would produce no additional costs above the baseline program of record in the sustainment and modernization programs.

Although all requirements for sustainment and modernization may not have been fully funded in the USAF POM or budget, these requirements should not change due to a decision to stop production after Lot 9.

However, Shutdown and Restart, Warm Production, and Continued Production would increase the costs of the F-22A sustainment and modernization programs. While in production hiatus, the sustainment contract work scope and funding would undoubtedly increase to cover personnel whose skills would be required for restarting production. This work scope and funding would decrease once production is restarted. A restarted production program may also lower unit costs for parts and equipment. Furthermore, since any post–Lot 9 F-22A aircraft procured are not in the baseline sustainment contract, they would increase the costs of the sustainment program.

Technical Data Package and Contract Closeout

The costs of procuring a technical data package are not included in the total cost of any option. Compared to the baseline program of record, the costs of the technical data package would be the same across all options, except for the effects of inflation.

Contract closeout costs would increase in proportion to the number of additional production contracts signed.

Industrial Base Implications

Our analysis of the F-22A industrial base suggests that while few vendors would respond to a two-year production hiatus by going out of business entirely, other issues are likely to hinder the program's restart capabilities. If restart in the future is desired, the vendor base must be closely managed, with actions taken to minimize the number of vendors requiring requalification. In addition, there are some concerns over the availability of skilled labor, processes, facilities, and tooling utilized by firms supporting F-22A production. DMS would persist during a hiatus and may worsen.

APPENDIX A

Vendor Assessments

Lockheed Martin and Boeing Vendor Assessment: November 20, 2007

This assessment seeks to gather information on vendors involved in F-22 production and to understand how slowed production and a production gap would likely impact their business activities.

Vendor Contribution

1. What proportion of your firm's costs for the F-22 are attributed to costs billed by vendors?
2. What proportion of vendor costs are attributed to single source suppliers?

Vendors Selected for Assessment

A two-tiered selection strategy will be used:

- Group 1: Vendors that were included in the prime contractors' response to RAND's initial survey. They include the top suppliers ranked in terms of their contribution to the material costs.
- Group 2: A random sample of 50 vendors selected for analysis from those not included in Group 1. RAND will draw the sample.

Vendor Information

General business information

1. Vendor name
2. Vendor address: city, state
3. Is the vendor a small, large, minority, veteran, or women-owned business?

Vendor Business Activities on F-22

1. What was the vendor's total contribution to the manufacturing cost of the most recent lot of F-22s?
2. Does the vendor have favored or preferential status? Do you have a long-term supply agreement with the vendor?
3. What products or services does the vendor provide for the F-22?
4. What systems in the F-22 do these products or services support?
5. Does the vendor provide goods or services for the F-22 modernization program?
6. How long has the vendor been supporting F-22 production?
7. Does the vendor have contracts with your firm to provide goods or services outside the F-22 program?
8. Is this vendor considered critical to F-22 production? If yes, please explain why.
9. Are the processes used by the vendor documented and available?
10. Do the products or services provided by this vendor utilize proprietary processes?
11. Does this vendor utilize employees with special labor skills that require extensive training or experience specific to the F-22?
12. Does this vendor utilize special facilities or tooling specific to the F-22?
13. Is the vendor an "only source" vendor?

14. Does the vendor generate a majority of its revenues from providing products or services for the F-22 program?
15. Do you anticipate that this vendor will undergo a product redesign within the next three years?
16. Assuming production continues at a rate of 20 F-22s per year, is this vendor at high, medium, or low risk of being unavailable after Lot 9 to support F-22 production?
17. Assuming production slows to five F-22s a year for two years after Lot 9, is this vendor at high, medium, or low risk of being unavailable to support F-22 production?
18. Is this vendor at high, medium, or low risk of being unavailable for F-22 production following a two-year production gap after Lot 9?
19. Will this vendor likely need to be requalified following a two-year production gap?
20. Is this vendor supporting production of the F-35?

Pratt & Whitney Vendor Assessment: November 7, 2007

This assessment seeks to gather information on vendors involved in F119 engine production and to understand how slowed production and a production gap would likely impact their business activities.

Vendor Contribution

1. What proportion of the total cost of the F119 engine is attributed to costs billed by vendors?
2. What proportions of vendor costs fall under a long-term agreement?

Vendors Selected for Assessment

The assessment should cover all first-tier vendors involved in production of the F119.

Vendor Information

General business information

1. Vendor name
2. Vendor address: city, state
3. Is the vendor a small, large, minority, veteran, or women-owned business?

Vendor Business Activities on the F-22

1. What was the vendor's total contribution to the manufacturing cost of the most recent lot of F119s?
2. Does the vendor have favored or preferential status? Do you have a long-term supply agreement with the vendor?
3. What products or services does the vendor provide for the F119?
4. What systems in the F119 do these products or services support?
5. Does the vendor provide goods or services for the F119 modernization program?
6. How long has the vendor been supporting the F119 production?
7. Does the vendor have contracts with your firm to provide goods or services outside the F119 program?
8. Is this vendor considered critical to F119 production? If yes, please explain why.
9. Are the processes used by the vendor documented and available?
10. Do the products or services provided by this vendor utilize proprietary processes?
11. Does this vendor utilize employees with special labor skills that require extensive training or experience specific to the F119?
12. Does this vendor utilize special facilities or tooling specific to the F119?
13. Is the vendor an "only source" vendor?
14. Does the vendor generate a majority of its revenues from providing products or services for the F119 program?

15. Do you anticipate that this vendor will undergo a product redesign within the next three years?
16. Assuming production continues at a rate of 20 F-22s per year, is this vendor at high, medium, or low risk of being unavailable after Lot 9 to support F119 production?
17. Assuming production slows to five F-22s a year for two years after Lot 9, is this vendor at high, medium, or low risk of being unavailable to support F119 production?
18. Is this vendor at high, medium, or low risk of being unavailable for F119 production following a two-year production gap after Lot 9?
19. Will this vendor likely need to be requalified following a two-year production gap?
20. Is this vendor supporting production of the F-35?

APPENDIX B

Previous RAND Research on Shutdown and Restart

Military cost analysis relies heavily on lessons learned and data analysis from past programs. A potential F-22 restart provides a unique challenge because no complex fourth-generation aircraft or beyond has resumed production after a gap. The manufacturing base needed for such a program is significantly more sophisticated than that required by most previous aviation systems: The tiers of specialized suppliers working with the three prime contractors make the F-22 assembly distinctly different from past programs with a restart history. Despite the differences, previous restart studies still provide useful information regarding consequences of system change. With respect to a manufacturing restart in the military aviation industry, two RAND studies (Birkler et al., 1993, and Younossi et al., 2001) offer useful lessons learned and post-production assessments.

Birkler et. al. (1993) obtained data regarding actual and proposed/planned restarts for seven military fixed-wing aircraft (B-1, C-5, C-140 Jet Star, F-117, OV-10, S-3 [Lockheed and LTV Aerospace, separately], and U-2), one commercial fixed-wing aircraft (B-707), two rotary wing aircraft (CH-46 and SH-60 LAMPS), and one air-to-ground missile (AGM-65). Extensive narrative data were obtained for all programs and most programs provided quantitative manufacturing labor data. Birkler's team was able to catalog general restart considerations, possible candidates for restart, general issues to consider, and—most notably—loss of learning associated with production gaps.

The feasibility and necessity of logistical planning needed to resume production directly impact cost. While all terminated production lines can be restarted, if required, the associated nonrecurring costs may prove prohibitive depending on run size. Birkler et al. note the importance of including the possibility of a production gap and restart as a part of normal procurement policy. They developed four sequential criteria for a program to be considered desirable for restart: (a) review cannot be postponed, (b) future demand is likely to be higher than current programmed levels, (c) another system cannot be substituted, and (d) restart is practical. The first three criteria were applied to an array of then-existing Army, Navy, and Air Force systems.

Birkler et al.'s analysis of restarted programs revealed several confounding factors that might arise during program gaps. These include changes to the government and contractor environments, the military and civilian regulations, and the contractual environment—accounting systems, organizational structure, and manufacturing processes. While these components arise exogenously to the aircraft program, they can greatly alter the manufacturers' operating space. The chance of confronting these issues increases with the length of the production gap.

Although the break duration may alter the operational environment, Birkler et al. found that it had no effect on the recurring labor hours post gap. Post-gap labor hours reflect loss of learning in the labor pool and may include some training and proficiency losses.

Loss of learning contributes a large portion of the aircraft recurring costs after a gap. Birkler et al. showed that, after a shutdown, the labor hour improvement curve shifts higher toward the level of the first pre-gap aircraft and eventually flattens asymptotically, thus barring a return to pre-gap labor hour improvements. The amount of production learning retained after the restart, however, varies significantly across programs.

Birkler et al. developed a heuristic about learning: The learning slope after a production gap (say 0.95) is the square root of the pre-gap value (say 0.9). It is important to note that this trend is highly variable depending on the aircraft and its manufacturing processes.

Younossi et al. (2001) applied these lessons learned to characterize the effects of a production gap for the U.S. Navy's E-2C. They partitioned effects into three areas: recurring cost (changes in the learning curve), nonrecurring cost (retention of critical workforce skills, facility planning, and supplier management), and industrial base sustainability during a production gap. They developed an analytical model to forecast the prime contractor's workload, productivity, and labor costs by labor trade.

APPENDIX C

Original and Restart Learning Curve Slopes

The restart production cost analysis in Chapter Three is focused on Equation 3.2, which generates the lot cost of aircraft. Equation C.1, the relation of retained unit percentage (RUP) and the pre-shutdown cost improvement slope (b) to the post-gap cost improvement slope (β), is the primary support to Equation 3.2. It requires further explanation:

$$\ln(\beta) = \ln(b) \times (-1.33949 \times \text{RUP} + 0.715537). \qquad \text{(C.1)}$$

The need for this equation is data-generated. Historical data showed that the historical fraction of units retained is highly correlated with steeper slope after restart.

This equation was estimated by looking at the production data from seven historical programs: Jetstar, U-2, LAMPS, CH-46, C-5, and S-3 (Lockheed), and S-3 (LTV). By fitting cost improvement curves, Birkler et. al. (1993, Appendix E) independently estimated the pre-gap and post-gap cost improvement slopes (here b and β). Once these were derived, they calculated the "quantity at which the unit cost on the original curve is equal to the first-unit cost of the restart curve," or $Q(OT_q = RT_1)$. For our analysis, RUP was calculated by dividing $Q(OT_q = RT_1)$ by the total units produced before shutdown. The parameters were estimated by applying ordinary least squares to Equation C.2, a modified form of Equation C.1:

$$\ln(\beta/\ln(b)) = c + d \times \text{RUP}. \qquad \text{(C.2)}$$

Parameters c and d are 0.715537 and –1.33949 and have standard errors of 0.09242 and 0.74583, respectively. The equation has an R-squared of 0.39.

Bibliography

Arena, Mark V., Obaid Younossi, Kevin Brancato, Irv Blickstein, and Clifford A. Grammich, *Why Has the Cost of Fixed-Wing Aircraft Risen? A Macroscopic Examination of the Trends in U.S. Military Aircraft Costs over the Past Several Decades*, Santa Monica, Calif.: RAND Corporation, MG-696-Navy/AF, 2008. As of December 10, 2008:
http://www.rand.org/pubs/monographs/MG696/

Bennett, John T., "Japan Defense Chief Asks for Secret F-22 Data," *Air Force Times*, May 2, 2007.

Birkler, John, Joseph P. Large, Giles K. Smith, and Fred Timson, *Reconstituting a Production Capability: Past Experience, Restart Criteria and Suggested Policies*, Santa Monica, Calif.: RAND Corporation, MR-273-ACQ, 1993. As of December 10, 2008:
http://www.rand.org/pubs/monograph_reports/MR273/

DoD Financial Management Regulation 7000.14-R, Vol. 2A, Ch. 1, Sec. 25, "Expired Appropriation," October 2008.

F-14 Shutdown Contract, Supplemental Agreement with Grumman Aerospace Corporation, Contract Number N00019-88-C-0276, Procurement Request Number 97-B7-TM008, Modification Number P0020, Naval Air Systems Command, AIR-21123, December 12, 1996.

Federal Acquisition Regulation (FAR), Subpart 4.804, June 14, 2007. As of December 4, 2008:
http://www.arnet.gov/far/97/html/04.html

Fisher, Gene H., *Cost Considerations in Systems Analysis*, Santa Monica, Calif.: RAND Corporation, R-490-ASD, 1970. As of December 10, 2008:
http://www.rand.org/pubs/reports/R0490/

GAO—*See* U.S. Government Accountability Office.

Goldberg, Matthew S., and Anduin E. Touw, *Statistical Methods for Learning Curves and Cost Analysis,* Linthicum, Md.: INFORMS, 2003.

King, D., and J. Driessnack, "Analysis of Competition in the Defense Industrial Base: An F-22 Case Study," *Contemporary Economic Policy*, Vol. 5, No. 1, January 1, 2007, pp. 57–66.

Land, Gerry, *Budget Exhibits*, Teaching Note, Defense Acquisition University, Business, Cost Estimating and Financial Management Department, April 2006. As of December 9, 2008:
https://acc.dau.mil/GetAttachment.aspx?id=30695&pname=file&lang=enUS&aid=5471

Tirpak, John A., "More F-22s, But Not 381," airforce-magazine.com, July 23, 2008. As of December 16, 2008:
http://www.airforce-magazine.com/DRArchive/Pages/2008/July%202008/July%2023%202008/MoreF-22sButNot381.aspx

U.S. Air Force, "F-22 RAPTOR," Air Force Fact Sheet, April 2008. As of December 4, 2008:
http://www.af.mil/information/factsheets/factsheet.asp?fsID=199

U.S. Government Accountability Office, "Subject: Tactical Aircraft: DOD Should Present a New F-22 Business Case Before Making Further Investments," June 20, 2006. As of December 4, 2008:
http://www.gao.gov/new.items/d06455r.pdf

Walker, David M., "Tactical Aircraft: Questions Concerning the F-22A's Business Case, Statement of David M. Walker, Comptroller General of the United States, Testimony Before the Subcommittee on Airland, Committee on Armed Services, United States Senate," July 25, 2006. As of May 22, 2008:
http://www.gao.gov/new.items/d06991t.pdf

Wynne, Michael W., "F-22A Multi Year Procurement: Statement of the Honorable Michael W. Wynne, Secretary of the Air Force, Department of the Air Force Presentation to the Committee on Armed Services Airland Subcommittee, United States Senate," July 25, 2006. As of May 22, 2008:
http://integrator.hanscom.af.mil/2006/August/08172006/072506Wynne.pdf

Younossi, Obaid, Mark V. Arena, Kevin Brancato, John C. Graser, Benjamin W. Goldsmith, Mark A. Lorell, Fred Timson, and Jerry M. Sollinger, *F-22A Multiyear Procurement Program: An Assessment of Cost Savings*, Santa Monica, Calif.: RAND Corporation, MG-664-OSD, 2007. As of December 10, 2008:
http://www.rand.org/pubs/monographs/MG664/

Younossi, Obaid, Mark V. Arena, Cynthia R. Cook, John C. Graser, Jon Grossman, and Mark Lorell, *The E-2C Industrial Base: Implications of a Production Hiatus*, Santa Monica, Calif.: RAND Corporation, 2001. Not available to the general public.

Younossi, Obaid, Kevin Brancato, Fred Timson, and Jerry Sollinger, *Starting Over: Technical, Schedule, and Cost Issues Involved with Restarting C-2 Production*, Santa Monica, Calif.: RAND Corporation, 2004. Not available to the general public.